Advance Praise for *Parent Nation*

"A manifesto, and a handbook, for what we as individuals and as a society are morally called to do for all kids to thrive. Required reading for anyone who has ever loved a child."

—Angela Duckworth, professor of psychology at University of
Pennsylvania and *New York Times* bestselling author of *Grit*

"*Parent Nation* makes clear that a child's first three years of life are a time of unparalleled brain growth, and our public policies need to reflect that reality. Societal supports for parents that begin on a child's first day of life, when learning begins, are the missing components in our nation's education system. Dr. Suskind's sensible and actionable recommendations illuminate our path to a future where all children are given a sturdy foundation and an opportunity to meet their inherent potential."

—Arne Duncan, former U.S. Secretary of Education
and author of *How Schools Work*

"A powerful reminder that we know too much about the critical importance of early brain development to continue treating children's first years as anything other than a paramount driver of health in our communities. Supporting children means supporting the adults that care for them; calling for and investing in coordinated, quality early childhood systems; and elevating businesses and governments that provide these vital structures— that's the work of all of us."

—Steve Nash, former NBA All-Star and MVP; coach of the Brooklyn Nets;
and president of the Steve Nash Foundation: Growing Health in Kids

"In *Parent Nation*, Dana Suskind compellingly argues that supporting parents as they raise young children should be a national priority. With empathy and a sense of urgency, Suskind movingly lays out why the neuroscience of early childhood development makes those years so critical. *Parent Nation* is nothing less than a call for action."

—Alex Kotlowitz, bestselling author of *There Are No Children Here*

PARENT NATION

Unlocking Every Child's Potential,

Fulfilling Society's Promise

DANA SUSKIND, MD

WITH LYDIA DENWORTH

DUTTON

DUTTON
An imprint of Penguin Random House LLC
penguinrandomhouse.com

LIBRARY OF CONGRESS CATALOGING-IN-PUBLICATION DATA
Names: Suskind, Dana, author. | Denworth, Lydia, 1966– author.
Title: Parent nation: unlocking every child's potential, fulfilling
society's promise / Dana Suskind, M.D., with Lydia Denworth.
Description: New York: Dutton, [2022] | Includes bibliographical
references and index.
Identifiers: LCCN 2021050017 (print) | LCCN 2021050018 (ebook) |
ISBN 9780593185605 (hardcover) | ISBN 9780593185612 (ebook)
Subjects: LCSH: Parenting. | Child development. | Child psychology.
Classification: LCC HQ755.8 .S88 2022 (print) | LCC HQ755.8 (ebook) |
DDC 649/.1—dc23/eng/20211026
LC record available at https://lccn.loc.gov/2021050017
LC ebook record available at https://lccn.loc.gov/2021050018

Printed in the United States of America
1st Printing

BOOK DESIGN BY PAULINE NEUWIRTH

While the author has made every effort to provide accurate telephone numbers,
internet addresses, and other contact information at the time of publication, neither the publisher
nor the author assumes any responsibility for errors or for changes that occur after publication.
Further, the publisher does not have any control over and does not assume any responsibility
for author or third-party websites or their content.

Some names and identifying characteristics have been changed to protect
the privacy of the individuals involved.

For my parents, Leslie Lewinter-Suskind and Robert Suskind,

*Whose love ensured that I would always see the beauty
and potential of humanity.*

D.S.

For Jacob, Matthew, and Alex,

Who made me a parent.

L.D.

• CONTENTS •

PARENT NATION

• AUTHOR'S NOTE •

Throughout this book, I use the word "parent" many times. Sometimes for variety and sometimes because the distinctions matter, I also refer to mothers, fathers, grandparents, caregivers, childcare providers, and other adults. I would like to emphasize that parents come in many forms and that what I mean by "parent" encompasses the broadest possible interpretation: a caring adult entrusted with the raising of a child. A parent nation, as I see it, is a society that cherishes and supports the love and labor that go into nurturing, raising, and educating future generations.

The parents you will meet in these pages are real, but I have taken some steps to protect their privacy. I have used only first names for the families I met through the TMW Center for Early Learning + Public Health. I did the same for the other parents I interviewed with a few exceptions, whose full names are included because their professional affiliations make them easily identifiable and are relevant. The names of Jade and her family, Justin, Katherine, and Ellen Clarke's friends are pseudonyms.

• PART ONE •

FOUNDATIONS

ONE

TOWARD A NEW NORTH STAR

"There can be no keener revelation of a society's soul than the way in which it treats its children."

—NELSON MANDELA[1]

As we near the "red line," the demarcation between the hospital's pre-op area and its collection of operating rooms, a mother and father hand me their baby. Their eyes are filled with tears as they look at me with a combination of hope and fear. The little boy is just eight months old and was born deaf. He is here to receive his cochlear implant. When I surgically implant the small device that will give him access to sound, I am replicating what I did for his father many years earlier when he was a teenager. As the baby melts into my arms, I reassure his nervous parents, "I promise to care for your baby like he's my own."

The parents settle in for a long, anxious wait, while I carry their son to the operating room. In OR4, where I spend each Tuesday morning, we are greeted by the team of medical professionals I rely on for every surgery, and by the cacophony of monitor beeps that I find so comforting every time I hear it. My two OR nurses are circulating. Gary Rogers makes sure the cochlear implant is present and that my favorite drill and facial nerve monitor are working properly. Nelson Floresco checks out the operating room microscope, which is the size of a Smart Car and gives me a remarkably clear, precise view of the ear's tiny, delicate interior spaces. Robin Mills, the OR tech, is scrubbed in and organizing the array of sterile microscopic ear instruments on the surgical table. The pediatric anesthesiologist gently places a face mask filled with colorless gases on the squirming baby. Very quickly, the baby is fast asleep.

Before I start the operation, we double-check that everything is in order. Do we have the right patient? Check. Do we have the correct implant with all the right instruments? Check. Do we know if the patient has any medical allergies? Check. Are the pre-operative antibiotics in? Check. This routine ensures the accuracy and safety of what we're doing. Each person in the operating room plays an essential role. No one forgets why we are here: to help a child.

As a surgeon performing delicate work just millimeters from the brain, I have no room for error. It's critical that I have the necessary tools and, even more importantly, my A-team by my side. If any part of this carefully crafted system falls away, no matter my skill or good intentions, my job will be infinitely more difficult. Some obstacles can be overcome—a few missing instruments, for instance. But what if the power went out in the hospital and I suddenly had to operate without light or oxygen? Or what if Robin, Gary, and Nelson suddenly walked out the door, leaving me alone? The odds would be stacked against me, and the job would seem impossible.

The challenge of successfully rearing children is not so different. To raise a child into a happy, healthy adult capable of achieving their full potential, you need a plan, and you need an appropriate, safe environment, one that provides backup as required. But far too many parents are not operating—that is, parenting—in an optimal environment. For too many parents, in our country and throughout the world, it is as if they are trying to function in the midst of an endless power failure, asked to achieve a critical goal without the necessary tools or any backup.

Twenty years ago, I started my own life as a parent with what I thought were all the necessary tools in place. But, in one painful day, it all changed, when my husband, Don, drowned while trying to rescue two boys, leaving me a young widow and my three children with no father. While we still had a roof over our heads and food on the table, advantages that many families lack, Don's death left a vast hole in our lives.

For a long time after he died, I would wake at night, jolted by the same terrifying nightmare, which went something like this: I'm standing on a foggy riverbank. Splinters of moonlight stream through the clouds and illuminate a small wooden boat next to me at the water's edge. Three small, terrified faces—my young children, Genevieve, Asher, and Amelie—peer from the boat, staring at the foreboding river. Its fierce currents resemble the waters of Lake Michigan, whose undertow claimed Don's life. I feel the intense pull of the water, the same pull that Don must have faced when he left the protective shoreline to swim toward the cries of the two struggling boys. Like Don, I have a desperate need to ensure young children are safe. In my dream, I have to get my kids across the river. I believe that if I can just do that, they will be okay . . . It will all be okay. But the torrent is too rough, the boat too flimsy, the opposite riverbank too far away. I wake sobbing, helpless, alone.

It was not hard to grasp the significance of my dream. I wanted what all parents want: to ferry my children into healthy, stable, and productive adult lives—that is what awaited on the far shore. I wanted to give them every opportunity. But it would take some time before I saw how fully all the elements of my dream—the turbulent water, the inadequate boat, the fact that there was no one standing next to me on that riverbank—symbolized the hurdles that so many parents face in the effort to successfully rear their children.

How could I navigate that torrent on my own, with no support, no help? How can anyone? Although I had been a surgeon for years and thought I had a deep familiarity with the lives of the families whose children I cared for, my struggles as a grieving, single parent gave me a new window of understanding into the challenges facing families.

Thirty Million Words . . . and Beyond

I became a surgeon because I thought I could change lives, one child at a time. By giving deaf children cochlear implants, I give them access to sound, to hearing, and to spoken language. I want there to be no barriers to their success, and I believe restoring their access to sound accomplishes that. Sign language can provide a rich, early language environment when provided by fluent signers. The baby whose surgery I just described is now fluent in two languages— American Sign Language and English. But the reality is that more than 90 percent of deaf children are born to hearing parents who don't sign. And early in my practice, I noticed profound differences in my patients' progress after surgery. Some children excelled developmentally, others not at all. Some learned to talk, others did not. The ability to hear, it turned out, did not always unlock their full capacity to learn and thrive intellectually. I could neither accept nor ignore the disturbing disparities I saw among my patients, but I didn't understand them. Compelled to discover their cause and to find solutions, I began a journey far outside the operating room and into the world of social science.

Initially, I was inspired by pioneering research that found a stark difference in the amount of language—the actual number of words—that children were exposed to early in life.[2] That difference often, although not always, fell along socioeconomic lines, with more language occurring in more affluent homes and less language in homes where families have been denied access to educational opportunities, often for generations. Researchers calculated that by the time children reached their fourth birthday, there was a gap of roughly thirty million words between those who heard a lot of language and those who heard very little. Although the research was done in hearing children, it explained what I was observing in my patients. In order to fully benefit from their new cochlear implants, they needed to hear a lively stream of words every day, they needed to practice listening. The quantity and quality of the words children hear stimulate the brain. Regions that are learning to process grammar and meaning will be critical to the ability to speak and later to read. Exposure to words also affects areas of the brain that handle feelings and reason, which will help children regulate their emotions and behavior as they grow older. The more language a child hears in those early years, the more securely the foundational connections are built in that child's brain.[3]

Some of my patients were getting that kind of essential experience with language, others were not. As I learned more, I realized that what I was seeing in my deaf patients mirrored the population at large and that this phenomenon was the basis of what is called the educational opportunity gap. In all children, the difference in early language exposure correlates with later differences in achievement. Too often that opportunity gap results in disparities between rich and poor children.[4]

The research was inspiring because it was based on the idea that parents are their children's first brain architects—that every parent, through the power of their words, has the ability to build their child's brain—and that we, therefore, have to make sure that parents have the resources they need to do that. The research also accentuated the urgency of actively building the brain during the first three years of life. Those early studies weren't perfect, and their limitations became clearer over time, but I think of them now as the first sentence in what has become an extensive body of literature.[5]

The work gave me and my colleagues a relatively simple explanation of the underlying disparities to target. It gave me a critical place to start. And it was persuasive enough to pull me out of the operating room for much of each week and into the world of research and social science.

In 2010, when I launched the Thirty Million Words Initiative, now the TMW Center for Early Learning + Public Health, my primary goal was to help ensure healthy development in all children and to give every child the ability to reach his or her potential, intellectually and emotionally.[6] Brain science pointed the way. Everything we designed and did was based on the fact that nurturing talk and interaction between caregivers and infants lay the foundations for brain development. My team and I developed evidence-based strategies to show parents the importance of talking to babies and young children. Those strategies became the theme of TMW: Tune In, Talk More, and Take Turns, or what we call the 3Ts. Our work is centered on the knowledge that rich conversation is what unlocks a child's potential and on the belief that parents as well as other loving caregivers hold the key during those early years. All adults—no matter their level of education, wealth, or work—can master the essential techniques for optimally building a child's brain.

The idea, a straightforward approach to a complex problem, was intuitively appealing and a great success. It was the "magic bullet" that people were looking for and it took me to the nation's capital, where I convened the first Washington, DC, conference on closing the word gap in 2013. Soon after, in 2015, I wrote a book called *Thirty Million Words: Building a Child's Brain*, which explained what research has revealed about the role of early language exposure in the development of children's brains. It was never just about the sheer number of words; but the difference between the effects of a lot of language exposure and a little served as a memorable representation of the brain-building strength of talk and interaction. The book caught on around the world. Everyone seemed to get its message. No matter the nuances of culture, vocabulary, or socio-economic status, people had an almost instinctive understanding of language as the key to developing the brain to its maximum potential.

Yet the more deeply I engaged in this new work, the more troubled I became. Or, to put it more honestly, the more I came to realize how naive my ideas were, limited by my own comfortable life circumstances. I had thought the answers lay in the actions and beliefs of individual parents, in their knowledge and behavior. (I still believe those elements are critical!) And it followed that the goal should be to ensure, as I put it in *Thirty Million Words*, that "all parents, everywhere, understood that a word spoken to a young child is not simply a word but a building block for that child's brain, nurturing a stable, empathetic, intelligent adult."[7] To that end, we were testing early language programs in randomized controlled trials—the scientific gold standard for determining what works and what doesn't. We found that, indeed, our strategies worked and the science that supports them is solid. The programs we promote at TMW can—and often do—improve the lives of children.[8]

But there was more to it than that. For our studies we recruited families, most of them low-income, from all over Chicago and later in other parts of the country. Our research followed children from their first day of life into kindergarten, and our programs took us into families' homes and into their lives. I was getting to know people up close and over time. The parents' enthusiasm was thrilling. They embraced the 3Ts with gusto, tuning in to their children, talking more as they went about their daily lives, and taking turns, encouraging their children to join the conversation. They wanted what we all want: to help their children get off to the best possible start. The problem was that the 3Ts took parents only so far. Real life would intrude, again and again and again.

There was Randy, who was excited to discover that talking about his love of baseball (Cubs only, never White Sox!) could help his son learn math but who had to work two jobs and, most days, had less than thirty minutes to spend with his kids. There was Sabrina, who gave up a well-paying job to care for her husband when he got sick and whose family ended up spending over two years in a homeless shelter, where she raised her two children, the youngest still a baby, in a stressful and chaotic environment. Most searing of all was the story of Michael and Keyonna, whose son, Mikeyon, missed out on all his father had to teach him for the first five years of his life because Michael spent that time in prison waiting to be tried for a crime he didn't commit—not appealing or serving a sentence, mind you, just waiting for his case to be heard.

Parenting is not done in a vacuum. Our research could not be, either. The circumstances varied, but everywhere I looked I saw the hurdles looming in front of mothers and fathers. At TMW, we can share with parents the knowledge and skills that build their children's brains, but our programs do not substantially change the day-to-day lives of the parents who participate. The larger realities of a family's circumstances—their work constraints, economic stresses, and mental health as well as the injustices and bad luck they are subject to—all matter as much as the 3Ts for healthy brain development. They either allow for the brain-building power of talk to occur or, if they limit the opportunities for engaging in the 3Ts, they stifle it like weeds choking the growth of a garden. When I saw just how difficult parenting is in a country that does so little to support the ability of parents to facilitate healthy brain development, I knew I had to learn more. I hoped I could *do* more.

Mirroring a Larger Problem

Reflecting on what I was seeing, I began to look beyond my patients and families at TMW to the entirety of the more than sixty million parents in the United States who have children under eighteen.[9]

And I saw how, regardless of income level, parents are being side-lined by our country's lack of family-friendly policies. I don't mean to minimize the herculean struggles of poor families or to suggest that more affluent families face equivalent challenges but to point out that society has abdicated its responsibility for *all* families. With the exception of the top 1 percent, our society makes raising children hard for everyone—and impossible for some. Some problems are obvious, others are more insidious. How is it that we spend less money on early childhood care and education than any other developed nation? By the end of 2021, why was the United States still the only one of the thirty-eight countries within the Organisation for Economic Co-operation and Development (OECD), an international body that seeks both to measure and to stimulate economic progress among its members, not to mandate paid parental leave?[10] The fact is that the great majority of parents have to work. Yet we have a fragmented and overwhelmingly low-quality childcare system; approximately half of Americans live in so-called childcare deserts, and fewer than 10 percent of existing programs were judged high quality in a National Institute of Child Health and Human Development study.[11] Given that roughly twenty million working Americans have children under the age of six and that more than 70 percent of mothers were in the workforce in 2020, this means that many millions of parents do not have adequately nurturing childcare for their children during the formative early years.[12] For this we can thank our economy. Wages for the middle and lower classes have been stagnant for decades. "Innovative" disruption has affected everything from bookstores to taxicabs and has created employment practices directly antithetical to the needs of parents and their children. The net effect is to place a heavy hand on the scales toward what benefits employers and their shareholders and away from what benefits families. In the process, and as a direct result, inequality has dramatically increased.[13]

To stay afloat, some parents have to take on multiple minimum-wage jobs that leave them little time for their children; others have the kind of job that requires constant contact with the workplace, via phone and computer, days, nights, and weekends. Everyone ends up overworked, stressed, and disconnected from family life.

As I talked to more people—you'll hear many of their stories—I saw how thoroughly all of this limits parents' choices. No matter their political or religious orientation, employment, or educational status, they all seemed to be struggling. I met Kimberly, a pediatrician at a community health center whose daughter was born prematurely at twenty-seven weeks. Kimberly had to leave her new baby in the neonatal intensive care unit just two weeks after her birth in order to go back to work. Imagine the pain of that! But her family could not afford to live without her salary, and Kimberly's state did not mandate paid family and medical leave, nor did her job offer it. I listened to Jade, who is deeply religious and believes that a mother's place is in the home, as she explained through tears that a lack of health insurance and an inadequate family income sent her back to work at Starbucks after her kids were born, despite her dreams of staying home. And I could relate to Talia, who had two babies while earning a PhD in psychology but gave up a promising postdoctoral position when it became untenable to manage the demands of the job, the economics of childcare, and the needs of two children under four.

Despite a culture that champions "family values," our society is not centered on families. It is not built around programs and policies that protect or promote those values. Quite the opposite. We erect daunting barriers in the path of far too many mothers and fathers—from mundane issues like irregular work hours that complicate childcare to profound structural problems like systemic racism that hold back sizable portions of our population. All these barriers limit the time and energy parents can devote to the brain development of their children. The barriers are unsupportive of parents, and they are holding back our next generation.

There is an alarming disconnect between what we know about what children's brains need and what we have actually done to develop those brains. At the very moment when parents and children could most use help, and when that help would have an outsize influence on children's ultimate ability to learn and succeed by strengthening neural connections, society does nothing—or worse, makes things harder. When it comes to children, public attention and money have been focused on K–12 schooling. But supporting children only during these years means we have skipped over the earlier phase that is critical to laying the foundation for learning at the K–12 level. Our efforts come too late for many, who will have been left so far behind during that critical period that by the time they get to kindergarten, they may never be able to catch up. Even preschool for all, while important, is not early enough.

The first three years of life are when the brain is in its most rapid, most critical period of growth. Successful education is predicated on the ability to learn, and that ability is dependent on what happens long before a child sets foot in kindergarten or even preschool. During those early critical years, parents are left largely on their own. This is why, despite decades of effort, we have not moved the needle on educational outcomes or equity. In the OECD's 2018 international educational rankings, the United States ranks 38 out of 79 countries in math and 19 in science.[14] Among developed countries, ours is near the bottom of the pack. We are the richest country in the world, per capita, yet we have lost sight of what is required to give all children a strong start on the road to being productive adults.

A Crystallizing Moment

I was already contemplating these deep-seated problems when the COVID-19 pandemic shut down the country in March 2020. At the University of Chicago Medical Center, where I work, it was all-hands-on-deck. I spent hours screening patients, talking to and corresponding with frightened people, noting their symptoms, and advising them on whether to go to the hospital. When I was on call as an ear, nose, and throat surgeon, my medical specialty meant I was working on the areas of the body—the nose and mouth—where the risk of transmitting the virus was highest. (The first doctor to die of COVID in China was an otolaryngologist like me.)[15] One difficult day I treated a man in his early forties who could not breathe on his own and needed a tracheotomy. Normally, that's a routine procedure to provide a surgical airway, but during COVID, it became a high-stakes procedure that required me to call in two chief residents to help. The medical side of the experience was exponentially harder than usual because of the anxiety and extra protocols COVID brought, but the human side of it was harrowing. As I stared at the man's thin, wasting body, I could see only hints of the strong construction worker he had been just a few weeks before. I knew his mother had already died from COVID-19 and that his wife was also sick and hospitalized in another unit. I had to wonder who was taking care of their young children and what would become of this family, which was being torn apart by this terrible disease.

And then, on April 21, more than a month into the pandemic, I got a text from Nelson, one-third of my operating room A-team.

‖ Pls pray for Gary Rogers. He was intubated today.

I was so shocked I could barely breathe. Because of the pandemic, we weren't performing elective surgeries, so we hadn't seen each other in a few weeks. But Gary, tall and strong with a quick wit and quicker smile, had been a warm, steady, supremely capable part of my life for years. Both Gary and Nelson had been OR nurses at Comer Children's Hospital within the University of Chicago

Medical Center since it opened in 2005. It was while at his second job, caring for dialysis patients—work he took on to help his daughters pay for college—that Gary contracted COVID. As a fifty-eight-year-old Black man, he was in a demographic group that seemed to be at higher risk of serious illness. And I knew as well as anyone that at that point in the pandemic, as doctors were scrambling to understand how to treat this new disease, once someone required a ventilator, the prognosis wasn't promising. I feared Gary was going to die.

After more than a month in the intensive care unit and two weeks on a ventilator, Gary was left with generalized muscular atrophy and cardiomyopathy and had to spend several weeks in rehab before he was strong enough to go home—and ultimately to return to work. When we reunited in OR4 for our first cochlear implant surgery in the summer, I was flooded with relief to have Gary, Nelson, and Robin (who had had a milder case of coronavirus) back together.

For a time, I took comfort in the thought that at least children were relatively immune to the virus. Alas, that was wishful thinking. Some did get sick (especially once the Delta variant of the virus arrived), many lost parents and loved ones, and nearly all suffered terribly from the loss of in-person schooling. The effects of the pandemic on children are still being calculated as I write. But within all the trauma and hardship of the pandemic, a sliver of positive news emerged. Even in the face of the extraordinary stresses the pandemic created—in many cases precisely because of those stresses—many families reported spending more time together. That was certainly true for me. With my kids (now in high school and college) in the house all the time, we had more family dinners than we had had in years. Even for families who suffered job losses, the pandemic's social safety nets helped some to cushion the blow and allowed families to enjoy being together. In March 2020, Congress's first relief bill, the $2.2 trillion CARES Act, replaced lost income for many workers, even those without unemployment insurance. Several more relief bills followed. They included, among other things, more direct payments to families and increased child tax credits. According to a study of recipients of the first round of pandemic aid, many of those who got those checks reported more positive parent-child interactions than those who didn't get checks.[16] But here's the rub: Parents were able to engage in conversations with their kids, to be there for their children and nurture their young brain cells, because the world had just about come to a complete stop. That is not real life. And the family conversations sometimes came at the cost of paychecks and financial security. That is not tenable. Eventually, most of the parents who were working remotely would have to return to the office, at least part-time, and the parents who were out of work would find new jobs. They had to. What would happen to family time and parent-child interaction then?

We can no longer deny how thoroughly entangled our private family lives are with our economic lives. Parents cannot work if their children do not have a safe place to spend the day. In the pandemic, schools closed and our already inadequate childcare system all but disintegrated. Two-thirds of childcare centers were closed in April 2020 and one-third remained closed in April 2021.[17] Even the Federal Reserve began to worry that childcare might be the broken leg of the economic stool that would make it impossible for the country to right itself.[18]

Parents were left on their own. Anxious and exhausted, they were called on to manage every aspect of their children's lives—to be teachers, coaches, therapists, and camp counselors—all day every day for the better part of a year in many places, longer in others. Even among those who didn't lose their jobs, millions ended up quitting (mostly mothers) or cutting back on work hours (again mostly mothers).[19] Doing it all was unsustainable. The pandemic was like a powerful earthquake with lingering aftershocks that showed just how shaky our nation's infrastructure of support for parents and, therefore, for children really was.

COVID was a crystallizing moment for me. As I watched its effects reverberate through the long months of distancing and difficulty, I was reminded that extreme situations can be clarifying. They show you what works, they show you where the weaknesses are, they show you what really matters. You cannot push pause on the work in progress that is a child's brain. And the pandemic was a forceful reminder that no one is meant to parent entirely alone. It really was the worst-case scenario from OR4, as if the power had failed, there was no oxygen or light, *and* the A-team had left me. (Gary nearly did!)

The pandemic also made plain that our current approach to children and families is both shortsighted and expensive. There was already plenty of evidence of that before the pandemic if you looked for it. Not investing in early childhood is estimated to cost our country billions. There is a cost to children, a cost to parents, and a cost to society. Economist and Nobel Laureate James Heckman of the University of Chicago has calculated that investments in programs supporting children from birth to age five (even programs that are very expensive in the short term) deliver a 13 percent annual return to society through better education, health, social, and economic outcomes well into the adulthood of the children served.[20] A failure to invest, on the other hand, means society ends up losing money because, without the preemptive protection of strong early childhood development, it must ultimately spend more on such things as healthcare, remedial education, and the criminal justice system. In short: If we don't invest in children from the earliest days of their lives, we—and they—do not just lose out on reaping the rewards of that investment, we pay a severe penalty for our failure. Consider this: A much-cited report by ReadyNation found that the overall cost to society of childcare issues is $57 billion a year and that the direct cost to employers is $12.7 billion. It has also been estimated that if American women stayed in the workforce at a rate similar to Norway's, which has government-subsidized childcare, the United States could add $1.6 trillion to the GDP.[21] Parents cannot work if there is no one to care for their children.

The Myth of Going It Alone

How did we get to this wholly untenable situation, where each parent stands alone on the dark riverbank? Where the dangerous currents and inadequate boats of my dream are made manifest in the hard realities of daily life? Where each parent's shoulders sag under the weight of the load? Somehow the centrifugal force of our societal choices flings children and parents to the outer reaches of our priorities instead of putting them at the center.

A string of deliberate political decisions, sins of omission, and unintended consequences are to blame. But one consistent theme runs through the choices we have made as a society: the mythic idea of American individualism. The roots of this idea reach to the nation's founding, to the colonial settlers and western pioneers who *had* to go it alone. Tough and independent, they made their own way because there was no alternative. We have been celebrating them ever since, even though our circumstances today are very different. Individualism perpetuates going it alone as a virtuous ideal. Expecting societal help is seen as a form of weakness, an admission of failure. And since the ideal of individualism is bound up with our ideas about the sanctity of our right to make our own decisions about our families and how we want to parent, such support is deemed inimical to liberty and freedom. At least that is how the story goes.

A key element of such thinking is the concept of parental "choice," which has been held up as sacrosanct, as the source of all parental authority. Anything else is considered un-American. The result has been to convince parents that they should be able to shoulder the enormous responsibility of early childhood care, development, and education on their own without formal support. Indeed, they should *want* to do so, should see it as a manifestation of their freedom to make decisions about their family life without interference.

As parents, we (especially moms) have internalized this propaganda. Burdened by guilt, most are managing a delicate balancing act, struggling to make it work, yet forever feeling inadequate, unable to live up to the ideal we imagine we should achieve. Occasionally, we get glimpses of an alternate universe when one of our own escapes the madhouse of the United States to another, saner country and finds that it really doesn't have to be this way. The popularity of *Perfect Madness* by Judith Warner and *Bringing Up Bébé* by Pamela Druckerman, both bestselling books that note the ample state-financed resources for parenting in France, reveals a desire for things to be different. And they could be.

In many other countries, support for family and parenting is increasingly recognized as an important part of social policies and investment packages aimed at reducing poverty, decreasing inequality, and promoting positive parental and child well-being. UNICEF is advocating for at least six months of paid leave for all parents, safe and comfortable public and workplace locations for women to breastfeed, and universal access to quality, affordable childcare from birth to the first day of first grade.[22] But here in the United States, we seem to have bought into the status quo idea. That, and perhaps our personal sense of failure as parents, keeps us from demanding more support from society. We are convinced we should be able to do this on our own and feel guilty about asking for help. I see this among my fellow physicians, my patients and friends, and the TMW families. I see it on the left and on the right, among the affluent and the poor. Few are spared.

In reality, choice and individualism for parents are myths—

convenient for those who wish to abdicate responsibility for offering support, horribly inconvenient for those who buy into the myth and suffer as a result. "Individualism" in parenting is more fantasy than reality, and "choice" borders on being an outright falsehood, implying as it does the availability of multiple options. In truth, most parents have few options and therefore not much to choose from, so how can we call that "freedom of choice"? Without support, there is no such thing as true choice. And you know what? In real life, except in a pandemic, almost no one actually parents alone. The reason the proverb "It takes a village" resonates is because it is true. Caring for children with zero help or community support is practically unheard of. There have always been grandparents, and aunts and uncles, and older siblings. There have always been neighbors and friends. There have always been other parents. Even pioneers circled the wagons to keep one another safe. We have offered one another advice, babysitting, moral support, and commiseration. We have been in it together. But valuable though they are, these private sources of support are not enough. Support systems are wonderful, but publicly financed and society-wide supportive systems are critical. We need more, and we should expect more, of our society.

Our Guiding Stars

Today we are in the midst of a public health crisis—one that goes far beyond the pandemic and will long outlast it unless we do something about it. Unlike COVID-19, it's a problem for which there is no vaccine. The lifelong impacts of early brain development are an invisible fault line running through society, magnifying and threatening to make permanent the disheartening inequities we see in our world. Multiple interwoven power structures, of economics, class, and race, ignore or actively undermine the ability of millions of parents to provide the stimulating, language-rich early learning environments they so desperately want their children to have.

In other words, the disparities that plague our nation begin far earlier in a child's life than most people realize. We are suffering from an invisible epidemic in the form of unequal opportunities for the early brain development that all children need to achieve their innate promise.

Sometimes the enormity of this crisis, the same one that pulled me from the operating room, is overwhelming. I once again feel as I did in my old dream, that I am standing on the dark riverbank. That we all are. But I'm also reminded of the words of Dr. Martin Luther King Jr. in the midst of the battle over civil rights that gripped this country in the 1960s: "Only when it is dark enough can you see the stars."[23] And I do see. I see with clarity two separate but inextricably intertwined ideas that allow us to move forward.

First, science gives us a road map. Just as it tells us what to prioritize individually as parents, it can show us where to go societally as well. It can lay out the coordinates that will lead us toward healthy brain development for all children. That goal, laying the foundation for optimal brain development, should be our constant guide. It will keep us focused on where we want to go as we set out to transform our society into one that makes its future citizens its focal point.

The science of brain development tells us to begin when learning begins, not on the first day of school but on the first day of life. Even in the womb, babies learn to recognize their parents' voices.[24] Timing is everything. Neuroplasticity, the brain's incredible ability to organize itself by forming new neural connections throughout life, is at its peak between birth and the age of three. Brain circuits are a use-it-or-lose-it proposition. While our brains remain plastic throughout our lives, they will never be more so than in the magical and essential early years.[25] To capitalize on this time, the all-important first step is rich conversation. It is often called serve-and-return, the back-and-forth of parents interacting with their children. Talking, smiling, pointing, responding—that nurturing interaction is powerful enough to help children move forward and develop two critical sets of skills that will allow them to succeed in school and in life. It delivers cognitive skills, the kind found on intelligence and aptitude tests: reading and writing, numeracy, pattern recognition. And it builds noncognitive, or "soft," skills like grit and resilience. In other words, nurturing interaction builds the whole brain.[26]

Neuroscience shows us that environment matters, too. Stable, calm environments foster socioemotional skills and executive function; disruptive environments impede their development.[27] Our society robs too many families of the opportunity to provide healthy environments. Illness. Poverty. Homelessness. These afflictions and others can trigger instability, and the resulting toxic stress becomes a risk factor endangering healthy brain development. When the ultimate development of a child is hampered, we all lose. Our future society will be made up of the children being reared today; therefore society should be helping to lay the foundation for optimal development of all its children.

If the science of the brain is our road map, it is parents who do the steering. That is the second critical point. Parents are the captains of their families' ships, manning the helm. But every captain needs a crew. It is time to reject the myth of individualism as justification for failing to provide societal support. That makes about as much sense as my walking into OR4 without my A-team. Having that A-team there does not diminish my control of the room. When parents hand me their child at the red line, they know exactly who is holding the scalpel. They are also glad to know that I have backup. Working together with me at the helm, my team and I get the job done. Having backup doesn't make me any less a surgeon, just as living in a society with family-friendly supports doesn't make a mom or dad any less a parent or any less in charge. Parents need true choice. They need authority *and* backup.

Building a Parent Nation

That's why this book about the importance of foundational brain development is called *Parent Nation*. Parents are the guardians of our future well-being. They should be recognized as the guardians of our present as well. Mothers and fathers are ordinary people— not one is endowed with superpowers—yet they accomplish something extraordinary when they raise children successfully. Parents are the architects of their children's brains and thus also the architects of society's future. It is only when we create a movement to support parents on their journey that we as a society can support the needs of early childhood. Loving mothers and fathers do not need a PhD or expensive gadgets to do an excellent job at supporting early brain development and building our future citizens. They need easily acquired, basic knowledge about how best to foster critical neural connections. They need time with their children to nurture those connections. They need high-quality childcare that complements their efforts. They need to be able to provide children with stress-free homes. And they need support for this formative endeavor from employers, from communities, and from policy makers—that's who I mean by "society."

When I wrote my first book, I thought that just knowing and understanding, and having others know and understand, the powerful brain science would be enough to bring about meaningful change. I was wrong. Real, essential change will occur only when there is a concerted, collective, national effort to bring it about. What we need is to recognize that we can lighten the parenting load by sharing it, by demanding what we require, and by asking society to help. What we need is to see the power in coming together as parents and as a nation to help all children. What we need is to put children's brain development at the heart of our thinking and planning.

By giving children the opportunity to achieve whatever their natural gifts allow, we fulfill the promise of their promise. Everything we do that affects families must begin there. In essence, we must reverse the spin and set up a society that pushes our focus inward, to the children—and their caregivers—at the center. We need to change the way society views an entire segment of the population: parents. Not just low-income parents. All parents. And in turn we need to change the way parents view themselves and elevate their expectations of support.

But how do we do it? By lifting our voices as one. There are tens of millions of us. Together, we can fight for our needs and our children's needs—for high-quality childcare, paid family leave, a child allowance. We can fight to address childhood poverty. We can demand that prenatal and pediatric care be holistic and include information about brain development. We can call on employers to institute family-friendly policies that are also good for their bottom line. If we form a coalition of parents, we can work together for the changes we need.

To create fundamental change, to ameliorate society's most entrenched problems, we must help all Americans to see that healthy brain development should be the North Star that guides us to a more productive, just, and equitable society. Addressing the issues of children and, therefore, their families doesn't help only those individuals; it is a necessary piece of addressing civil rights, gender equality, and the strength of our economy. So far, we have failed to see it that way. The ramifications of that critical failure are becoming more impossible to ignore with each passing day and were brought into the highest relief during the pandemic.

As a physician caring for children for over twenty years, I can attest to the fact that there are no fiercer advocates for children than their parents. I've seen it, time and time again. It is a beautiful thing to behold. Parents want to give their children what is rightfully theirs, the promise of their promise, even in the face of extreme obstacles. What if we could harness that passion, persistence, and determination into a movement that would compel society to deliver on children's unalienable right to realize their potential? What if we could convince society to make foundational brain development our guiding principle, our new North Star?

The beauty of this approach lies in its capacity to benefit each and every one of us, even non-parents. Undoubtedly, it will help to level the playing field and ensure that all children have a better shot at reaching their full potential and matching the achievements of their peers. The fate of each child, no matter how well nurtured, is, ultimately, intimately intertwined with the fates of all children. The strength of our country is based on ensuring that all our children have the *same* opportunity.

Being a parent has the power to bring us to our knees. But what brings us to our knees must also rouse us to our feet. Change doesn't happen spontaneously. These days, I dream of parents lined up next to me on the shoreline, millions of us setting out together, with our children, in sturdy boats capable of navigating even the most torrential river. I hope this book will remind parents that there is more that unites us than separates us; that it will help parents see that they are not alone in their struggles or aspirations for their children; and that it will make clear to parents that we are stronger together. I hope it will give parents and their allies what they need to succeed. And that, together, we will build a parent nation.

TWO

THE BRAIN'S GREATEST TRICK

"I believe every person is born with talent."

—MAYA ANGELOU[1]

Despite spending decades operating so close to the human brain, I am routinely amazed—and I mean stop-dead-in-my-tracks amazed—by its complexities and capabilities. This was certainly the case when a paper by my mentor and colleague, Susan Levine, landed in my inbox a few years ago. Susan is an expert in both language development and cognitive development. This was a study of a teenage girl whom Susan and the other researchers called C1, but I will call Charlotte.[2]

Charlotte was born with a rare condition called hemihydranencephaly; more simply, she has half a brain. In utero, the blood supply to her left hemisphere had likely been cut off, preventing her brain from growing and developing normally. While the most basic, ancient structures of her brain—the ones responsible for involuntary actions like breathing and motor functions—were intact, her left hemisphere never grew. Instead, the space filled with cerebrospinal fluid. In a brain scan, there is a large black spot where half her brain ought to be. The regions Charlotte is missing are those typically responsible for logic, language, and reasoning. The prognosis for such a condition is obviously devastating.

For a moment, imagine your doctor informing you that your beautiful new child has been born with half a brain. All your hopes and dreams for her future would be replaced with shock, sorrow, and uncertainty. You would likely assume that if she survived, your daughter would have severe developmental delays—that her potential for living independently and flourishing in the world had been destroyed before she was even born.

But that is not at all what happened to Charlotte.

Susan and her colleagues have been following Charlotte from the time she was fourteen months old. The paper I read described the first fourteen years of Charlotte's life (she is now several years older). By her teens, she still lived with some mild motor weakness on her right side. But otherwise, she seemed unaffected. She finished high school and was headed off to college.

How is that possible?

The answer lies hidden in the same place as the problem: Charlotte's brain. Before I explain, let's consider the wonders not just of Charlotte's brain but of any child's.

The human brain, said Isaac Asimov, is just three pounds, but those three pounds are "far more complex than a star."[3] Its bumps and folds are fragile, no firmer than a tub of soft butter, but within them is an intricate, buzzing world, the command center for thinking and learning, for being human. For being *you*. The brain controls our breathing and heart rate. It helps us learn to speak and to understand language. It recognizes that other people have beliefs and emotions of their own. It leads us to feel anxious or assured, elated or despondent. It governs our ability to sit still or delay gratification. It gives us the capacity to read books and to write them, to count, multiply, and calculate differential equations, to see the

relevance of history and appreciate the gravity and grace of the Get-
tysburg Address or King's "I Have a Dream" speech. It wonders how
far away the stars are, why dogs and dolphins behave as they do,
what causes cancer, and even how the brain itself works. Then it
plans experiments to figure out the answers.

When a child is born, nearly everything on that long list of abilities lies ahead. A new brain is very much a work in progress. I remember gazing into the tiny, scrunched-up face of each of my
children right after they were born and wondering who they would
become. I was awed by the fact that a newborn infant is a bundle of
potential with the capability of developing into a particular person.
But potential is not certainty. The genetics a child inherits from its
mother and father lay down some possible plot lines, but they provide only a first and fuzzy draft of a life story. Neuroplasticity, the
brain's astonishing capacity to change with experience, has a big
hand in writing the book, too.

The brain's ability to change through its interactions with the
world, to rewire itself based on experience and adapt to its environment, is its greatest trick. It offers unimaginable opportunity but
also brings tremendous risk. Just as areas of the brain can be
strengthened and enhanced—which, as it turns out, is what happened for Charlotte—they can also be stunted and diminished.
Each brand-new baby brain contains billions of neurons, but there
is little communication between them. In the first months and years
of life, the number of neurons inside a young brain explodes. More
significantly, so do the new connections that form between brain
cells at an estimated rate of one million per second.[4] Every new
experience—what the baby hears, sees, touches, tastes, and smells,
each caress and snuggle, lullaby, or instruction—serves as a guide.
You can think of those experiences as information that the brain
uses to fine-tune its setup or revise its manuscript. Each piece of
information sparks electrical impulses that travel from one neuron

to the next, jumping over the small gaps between cells called synapses. When a series of neurons communicate regularly, they get into a routine, like confident and practiced dance partners who fall into an easy rhythm together and don't have to think about the next step. Communication—really electrochemical signaling—becomes streamlined and efficient. The process is summed up in the popular maxim: "Cells that fire together wire together."

As cells wire together, they create circuits, loops of connected neurons in various parts of the brain that underpin the abilities children go on to acquire, and the child's newfound skills in turn affect the development of these circuits. The first to form are basic sensory processing loops. As babies get better at identifying faces, for instance, especially the all-important ones belonging to their parents, the visual cortex in the occipital lobe at the back of the brain is wiring up. Then more complex circuits are built on top of that.

Since I am a cochlear implant surgeon, I'll use hearing to show how basic and more complex circuits are related and the role environmental input plays. Newborn hearing babies are surrounded by a stream of sound, and through the first year of life, one of their main jobs is to make sense of that stream, picking out patterns and recognizing sounds that are repeated. If Liz, a playful mother, says, "Peekaboo, baby Jack! I see you!" again and again, day after day, little Jack will soon recognize his own name, and, for good measure, he'll associate the word "peekaboo" with fun. Listening strengthens the circuits in the auditory cortex, the part of the brain that governs hearing. Eventually, as Jack learns to talk, the language areas of his brain, which sit in the temporal lobe just above his ear, will use the hearing circuits that have already been laid down and will add to them, pulling in motor areas so that he can physically form the sounds he has been hearing and express them himself. One day—it will seem sudden, though it's anything but—when his mother points to a golden retriever fetching a stick in the park and says, "See the doggie?" Jack will cry, "Doggie!" Later still, when Jack starts to de-code the symbols on the pages of books ("D is for dog"), his brain will create a reading circuit that draws on the language and hearing and vision he has already mastered. This process of generating new neurons and creating neural connections is what the learning of early childhood is all about.

Around the age of three, however, an important shift in the neu-ral action gains momentum and it helps explain why the first three years of life are so critical. The rapid growth in the number and especially the networking of brain cells is followed by a gradual but ruthless process of elimination, a pruning of the neural connections that didn't pan out or prove necessary.[5] The unused nerve fibers wither and are reabsorbed into the brain tissue. This paring back of neural connections may sound like a bad idea, but it serves an im-portant purpose. Pruning keeps our brains efficient and helps them focus on what's important. It brings order to what might otherwise be a chaotic and overwhelming scene. Anyone who has seen tod-dlers melt down at their own birthday parties (yes, Asher, I'm talking about you) knows that young brains need help introducing organization and keeping calm. With pruning, the circuits that re-main are those that are used regularly. They become sturdy and reliable brain architecture. That's why we want to repeatedly rein-force which connections are useful and worth keeping.

In effect, we want to give the pruning process a lot of material to work with. If important connections never form in the first place, there's no chance for them to be part of the final brain circuitry. Having too little input to stimulate their developing brains is why children born into less vibrant language environments don't develop the same strong foundations. I learned that it is also the key to the mystery of why some of my deaf patients didn't develop strong lan-guage skills after cochlear implantation.[6]

The biggest surge in the growth and then pruning of neural

connections that will happen in an entire life span occurs in the first few years of life. There will never be a more effective time to establish the foundation for learning and development. In the first thousand days of a child's life, over 85 percent of the brain's total adult volume is built, which is why what happens during this time is so critical.[7] Early experiences cast a long shadow. Scientists have shown that building sturdy brain architecture leads to stronger literary, reasoning, and other skills; it increases academic achievement and decreases the chances of dropping out of high school. Early childhood development also affects lifelong physical and mental health; it is associated with lower levels of obesity, type 2 diabetes, and heart disease, among other conditions.[8] Strong early childhood care and education are even linked to reductions in crime and increases in lifetime income.[9] When it comes to brain circuitry, it's better to get it right the first time than try to fix it later. Like creating a home on a reliable foundation, starting with a solid underpinning is essential to ensure that what is built on top does not wobble or weaken.

What about a baby like Charlotte, who had a brain with a physical structure that was so compromised? Amazingly, in the case of babies born with brain defects, neuroplasticity can help the parts of the brain that are healthy and intact to rally and pick up the neuronal slack. The half of the brain that Charlotte had learned to do double duty! By the time she was a teenager, her brain scans reveal that the white matter connectivity, which determines the efficiency of communication between brain cells, in her remaining right hemisphere was much stronger than in typical children her age.[10] Her brain has compensated; functions that might usually occur in the left hemisphere are delegated to the right. Which is why Charlotte was able to amaze everyone around her by accomplishing so much with the right hemisphere alone.

But that's not even the most astonishing part.

When researchers compared Charlotte's cognitive function with a control group of typically developing kids, they discovered that in most areas, she was keeping pace with her peers, and in some areas, she was actually surpassing them! That kind of accomplishment took time. Early on, Charlotte's expressive and receptive language (what she could say and understand) was quite delayed. But she began to read and decode words when she was only three years old. By middle school, although her vocabulary and reading comprehension were below average, most of her language skills were in the typical range. And some of them, such as her decoding and reasoning skills, were above average. The brain injury exerted a price, but it is amazing that she did so well in the face of it.

The twofold explanation for how this could happen, for how we can maximize the gift of neuroplasticity, is the key to unlocking the potential of *all* children, not just the ones born with brain injuries. The first essential factor is timing. Charlotte's brain injury happened while she was still in her mother's womb. As we've already seen, neuroplasticity, the brain's astonishing ability to reorganize itself by forming new neural connections throughout life, is at its peak between birth and the age of three. We know that when major brain injuries happen in older children and adults, the capacity of the brain to adjust and adapt is drastically reduced and outcomes are radically different.[11] If what happened to Charlotte's brain had occurred when she was an adult, or even a teenager, she might have died and would certainly have been gravely affected. But because it occurred when she was a baby, there was time for the brain's circuits to reroute themselves to use parts of the right hemisphere instead of the more usual left side of the brain.

The second reason Charlotte was able to triumph in spite of her condition was the environment in which she developed, from the very first day of life. She vividly illustrates what can happen when someone, even in an extremely precarious state, is born into an environment that supports their potential. I assume Charlotte's parents aimed a constant stream of words and baby talk at her even before they learned about her condition when their daughter was ten months old. Once they discovered what Charlotte was up against and were told how critical the early language environment would be to her chances of overcoming it, they became deeply engaged in helping her. They made sure to provide the plentiful amounts of stimulation and input—conversation, interaction, warm cuddles, and more—that were essential to her ultimate brain development and behavioral outcomes. The family also got help from professionals through early intervention services. Charlotte began physical and occupational therapy while still a baby and speech therapy as a toddler. The therapy she got complemented everything her parents were doing. Her brain spent those early years completing its rewiring. Despite her slow start in language, she had essentially caught up by the time she was four.

Charlotte is not the only person born with half a brain. Not every story turns out as well as hers, but a Google search will turn up a surprising number of other people who were unaware for years that their brains were not fully formed and who operated in the world with no sign of trouble. A 2007 report in *The Lancet* described a forty-four-year-old man who was leading a normal life but was missing the majority of his brain. The black gaps in his MRIs are even more dramatic than Charlotte's.[12]

Stories like these highlight a profound truth: What happens in the first three years of a child's life has lifelong consequences, for better or for worse. Those years represent an opportunity that won't come again.

Charlotte's story also highlights a second, more disturbing truth: There are millions upon millions of children born with fully functioning brains who are left by the wayside in our world. The insult to their brains does not come in utero but arrives shortly afterward, when society denies them the healthy experiences they need to meet their potential.

I've often wondered what would happen if the incredible neural connections that are formed every second in a young baby's brain were visible to the human eye. Would we then be more proactive in how we invest in our youngest citizens? Or, even more significantly, if we could see when they *weren't* occurring in millions of babies, would we, knowing the consequences, be spurred to act sooner, with more intent, with more fervor? Would we be more willing to invest as a society?

Neural changes are less tangible than the physical change in a child who one day snuggles in your lap and the next is too big to carry up the stairs. (Oh, how I remember exactly that moment!) Hidden away in the skull, the brain's growth and development occurs in its very own black box. We don't see brain connections strengthening the way we see muscles getting stronger and children getting taller. But it's happening. When children say their first word or are suddenly capable of putting three words, and then five, together in a sentence, or recognizing the letter *B*, their new capacities are thanks to the extraordinary growth happening inside their brains. If we don't support that growth and give children the inputs they need, they will be shortchanged of their birthright.

Invisible Epidemic

The brain doesn't know what income bracket or zip code it was born into, it just shows up—a fist-size world of possibility. But too often that income bracket or zip code will end up making a very big difference in how that brain is built and, therefore, in the trajectory of a child's life. Of all the circumstances that can influence what happens to a brain, one of the most deleterious is also one of the most common: poverty. When I began my journey to try to understand the disparities in children's development, I had no idea I was sitting on the edge of a modern-day epidemic, one that predated COVID-19. In the United States, the likelihood of a child being born poor or near poor, meaning into a family that lives at 200 percent of the poverty line or below, is 40 percent. That rises to more than half for children of color.[13] Researchers who study health disparities have already shown that being born poor is associated not just with staying poor throughout life but with myriad unhealthy outcomes such as higher likelihood of substance abuse and cardiovascular disease as well as shorter life span overall.[14] Poverty's effect on the brain is particularly insidious. Unlike malnutrition, which is easily observable in a tiny, frail infant, poverty's impact on the brain is hidden behind the chubby cheeks and soft skin of an adorable baby, only manifesting many years later.

Broadly speaking, we have known for some time that adversity, especially in the early years, can have a harmful effect on child development. But it was only about fifteen years ago that a few neuroscientists began to seriously explore how the well-known effects of socioeconomic inequality change the brain. One of those scientists was Kimberly Noble.

Like me, Kim is a doctor, a pediatrician. Also like me, Kim stepped outside the clinic walls into the world of her patients as a research scientist. We both believe the brain holds the answers to many of society's ills and is the key to giving every child a fair chance. There are so many parallels in our work, in fact, that the first time we met, when Kim came up to introduce herself at a conference, we hugged as if we were old friends. It felt like we were!

Kim went through her undergraduate years at the University of Pennsylvania thinking she would be a psychology professor, but she also worked for a neurologist who invited her along to meet patients.

The experience changed her life. Hooked on medicine, she changed her plans and went on to get a combined MD/PhD degree. A PhD requires original research, and Kim found her way into the laboratory of renowned neuroscientist Martha Farah, who was looking to apply neuroscience to real life. Together, Martha Farah and Kim Noble were among the first researchers to investigate the effects of poverty on brain architecture.

What they and others have found is alarming. While there is considerable individual variability, poverty is associated, on average, with clear changes in the brain that make an already hard road for children harder.[15] It can steal the promise of a child's promise.

But it doesn't have to be that way.

"We start out pretty much the same," Kim says. Although each child comes into the world with genetic differences that will establish their range of potential, there is no evidence of association between an infant's socioeconomic status and their brain wiring at birth. We know this because we can use techniques like electroencephalography or EEG to record and measure different types of brain waves related to brain functioning. It's like eavesdropping on the brain's conversation.[16] At birth, I think you'd be hard-pressed to distinguish between the EEG readouts of a baby born in the favelas of Rio de Janeiro and those of the baby of an investment banker who works on Wall Street. It is only later that any differences related to a child's socioeconomic status show up in EEGs. By six to twelve months, there are significant differences suggesting that the postnatal environment may play a very important role.

Some of those differences are structural, meaning they concern the anatomy of the brain—its physical size and shape. You can think of structure the way you think of the hardware in a computer—it encompasses the processing chip and the wiring circuits that must be built so that software can run. In a study of more than one thousand children between the ages of three and twenty, Kim and her colleagues looked carefully at very specific measures of brain structure. The cerebral cortex, the thin layer of cells on the outer surface of our brains "that does most of the heavy lifting cognitively," as Kim puts it, is three-dimensional. You can measure its volume, thickness, and surface area. Kim zeroed in on surface area, which includes every nook and cranny in the bumps and folds of the brain and progressively expands over the course of childhood. A larger surface area tends to be associated with higher cognitive ability, and Kim and her colleagues found a consistent relationship between cortical surface area and socioeconomic factors, specifically parents' education and family income.[17]

Other studies have found differences in how the brain functions. If the brain's structure is considered the hardware of a computer, brain function is what happens when you boot up a computer and ask it to search for a document or stream a video. Measuring brain function means assessing how exactly a brain does any particular task. Which parts of the brain are involved? Is it efficient? Can it do what's required? Poverty makes many tasks harder. Research shows that material hardship clearly affects several parts of brain function that will be critical in determining the child's trajectory in school: language, executive function, and memory.[18]

Language ability, as my own work has shown, is highly susceptible to what happens in the first years of life. We know that there are marked differences in the quantity and quality of language and conversation that children are exposed to in their day-to-day lives and that poorer linguistic environments in turn affect the circuitry that is being constructed in the brain's language areas. Those language-based differences ultimately affect learning capacity.[19]

Executive function and memory also appear to be strongly affected by early experiences. Both are supported by the same regions of the brain that are very sensitive to stress. Toxic levels of stress, which all too many poor children experience, affect the formation of three critical parts of the brain: the hippocampus, essential for memory; the amygdala, central to emotion; and the prefrontal cortex, critical for reasoning, judgment, and self-regulation.[20] As with limited language exposure, negative influences (like stress) on the development of these important brain areas lead to poorer academic outcomes.

But how does being born into poverty cause these changes? That is part of what scientists like Kim are trying to determine. Income and parental education are proxies for other things. Lower levels of either often mean that families live in neighborhoods with fewer resources, more pollution, less comprehensive healthcare, and higher crime, and that the parents work long hours with little time to spend with their children. As a result, factors like poor nutrition, exposure to toxins, lack of prenatal health, and limited cognitive stimulation are among the likely culprits behind the research findings that show young brains are harmed by poverty.[21]

The depressing result is that as early as nine months of age, infants born into poverty score lower on cognitive development tests than their more affluent peers. By kindergarten, children living in poverty are likely to have cognitive scores that are, on average, 60 percent lower than their more fortunate peers.[22] This is, simply put, an invisible epidemic. Children are suffering from disparities in the very development of their brains. The effects of these disparities are lifelong—negatively impacting everything from academic performance to physical and mental health to employment.

While, on average, all of the above is true of kids born into poverty, it is also true that there are many individuals who don't fit neatly into this frame. There is real variability: Many children from poorer families have larger cortical surface area, for example, and many more advantaged children have smaller total surface area. On the other hand, there is no question that the disadvantages disproportionately impact children born in poverty. The poorer you are, the worse the consequences for your brain. One of the more striking points in Kim's work was that the most damaging effects were seen at the lowest levels of income.[23] But there are encouraging recent results from the Baby's First Years Study, an ambitious and rigorous study by Kim and her colleagues in which some lower-income mothers were given a monthly stipend to reduce poverty. After the babies turned one, those whose mothers received the cash showed different and faster brain activity, the kind that is associated with stronger cognitive skills.[24]

What is alarming about all this is the fact that children make up the poorest segment of the American population.[25] Think about that. Appallingly large numbers of babies and young children are living in conditions we *know* to be harmful to their brains. Each of those children is starting out with daunting odds in the lottery of life. This certainly contradicts our idea of our country as a place where no matter your background, you can achieve anything if you just try hard enough.

The Lost Einsteins

"Talent is equally distributed, opportunity is not." That's a variation on a popular line that perfectly sums up the problem we face: unequal opportunity.[26] For a long time, the phrase "achievement gap" has been used to describe the disparity in academic outcomes between affluent students and those from lower-income families—many of them people of color, non–native English speakers, and those living in rural areas. But recently, educators and social scientists have shifted to using the phrase "opportunity gap," which more accurately describes what's going on in the country (and around the world). It captures the fact that it is the circumstances into which people are born that shape the opportunities they will go on to have in life. Far too often, those circumstances are the result of conditions that society sets up. Or, as some education experts have put it, the opportunity gap is "a systemic problem, not a kid problem."[27] When given resources and support, all kids can thrive and achieve their potential, whatever that may be.

There is another kind of gap that should worry us, too. It is sometimes called the "excellence gap." Remember when I said each child comes into the world with genetic differences that will establish their range of potential? We are not all the same, and the variability in natural ability is not tied to income. The phrase "excellence gap" describes a phenomenon in which children of high ability aren't given an equal shot in life.[28] There are a lot of kids like that. In an average year, according to an estimate by the Jack Kent Cooke Foundation, nearly three and a half million high-achieving, low-income kids fill American classrooms from kindergarten through high school. For such children the problem is not making sure they achieve minimum competency, it is making sure they flourish and walk through the doors their talent ought to open for them. Yet those three and a half million smart poor kids are less likely to take challenging classes, to apply for federal financial aid, or even apply to college.[29] A huge swath of children who ought to be among our most productive and prolific will be held back by the circumstances of their birth. Knowing these kids are out there, we should be doing everything in our power to help them succeed from the first days of life all the way through school. But while opportunity gaps have been holding steady or slightly narrowing, excellence gaps have been widening.[30] There are just too many barriers in the way. We are potentially losing world-changing human capital when we lose out on what these children have to offer. The brightest of these, when they don't have the opportunity to show the world what they're capable of, have been hauntingly called "lost Einsteins."

When Hazim Hardeman entered his senior year in high school in North Philadelphia, he was a C and D student. That's not surprising, considering he spent more of his school days in hallways and bathrooms, gambling, than in class. Outside school, he could usually be found on the basketball court at the local rec center. He thought basketball was his path to glory or at least to a way out of his

circumstances—he spent the early parts of his life in public housing, bouncing around and nearly failing mostly suboptimal schools
(with one notable exception, which I will come back to). So you
might be surprised when I ask you to hazard a guess at what Hazim
has in common with President Bill Clinton, Transportation Secretary Pete Buttigieg, former national security advisor and ambassa-
dor to the United Nations Susan Rice, and physician and bestselling
author Atul Gawande.

Before I tell you the answer, I want to tell you Hazim's story. Take
note of the many moments when his life could have taken a very
different turn, one from which there would have been no recover-
ing. The times when Hazim almost became a lost Einstein. Rather
than an "up-by-your-bootstraps" sort of story, his could just as well
have been a cautionary tale.

Hazim's mother, Gwendolyn, had beautiful aspirations for him.
By the time Hazim was born in 1994, she had landed in Philadelphia, determined to give him and her three other children access to
the stable and loving home she'd been denied growing up in foster
care in Atlanta. That was no easy feat. A Black single mother, Gwendolyn had no extended family support network. Even though they
were poor, Gwendolyn prided herself on "having one of the best
houses in the projects," Hazim remembers. "Although we didn't
have an abundance of material resources, she always made us feel
like, if there was something we wanted, we would get it, even if she
couldn't get it immediately." That was especially true of education.
Gwendolyn's own struggles had shown her how life-changing it
would be to give her children access to a good education, and she
made it clear that school mattered, encouraging her children "to be
curious, to be critical and not to settle." In some ways, Hazim says,
his mother saw the education of her children as "a life-and-death
situation."

When Hazim was a child, doctors told Gwendolyn that he had

attention deficit hyperactivity disorder (ADHD). "I was always bouncing off the walls," he says. Hazim got medication and was assigned to a social worker. The diagnosis of ADHD might have ended up pushing Hazim into remedial classes and even expulsion, a well-worn path for so-called disruptive Black boys that begins as early as preschool.[31] Instead, as Hazim tells it, Gwendolyn decided he needed to change schools. (During one visit to Hazim's elementary school, she had seen students standing on desks, and teachers unable to control the classroom.) One day Gwendolyn took the bus to a distant, wealthier neighborhood, found what looked like the best K–8 school in the neighborhood, and, taking a big risk, because people have been jailed for this, she registered Hazim and his brother there, using a false address. She "literally put her freedom on the line to make sure we had a quality education," Hazim says. They spent three years there.

Immediately, the boys noticed stark differences between the new school and their previous one. The school in their neighborhood had primarily Black students. This new one had mostly white and Asian. The material differences were endless. "It just simply had resources— books, teachers who weren't overburdened, extracurricular activities," Hazim says, noting that such resources enrich both the teaching in the classroom and "the hidden curriculum." At their old school, it had felt like they constantly needed to prove they were smart and worthy of resources. At the new school, intelligence was assumed, and children were encouraged to explore and be curious. Hazim saw that his new classmates carried themselves differently, that they knew they belonged. He remembers wanting to belong, too, wanting to be treated as intelligent. Mostly, though, he managed the situation by being the class clown.

Listening to his story, I felt sure there must have been hints of Hazim's gifts early on—that is what we want to believe, that when the light is just right, it will hit the face of the diamond in the rough and glint so we will know that this is a person whose gifts must be nurtured. But when I asked him, Hazim insisted it wasn't like that. He was never better than an average student through all this time. In other words, even in this more promising setting, his potential remained hidden. The real value of his years in that school lay not in what he showed the world but in what it showed him, in what he saw was possible.

During high school, back at a poor-performing school in his own neighborhood, Hazim skipped class and played cards. Many of his friends were funneled into jail or died. Hazim suspects one reason he kept out of more serious trouble was because, for eighteen months, he lived with his older sister while his mother was in Atlanta caring for a family member. His sister's house was far enough away from his old haunts to limit his exposure to those friends. Still, there were some close calls, he said, "situations where I've turned left, and my friends have turned right." Once, as he was hanging out in his old neighborhood, cops showed up. They were looking for a robbery suspect. Hazim and the friend he was with had nothing to do with the robbery, but they ran anyway. The friend was caught, Hazim was not. Speaking of his friend, Hazim says, "I don't know if this was his first contact [with the criminal justice system], but I know he was very young, younger than I was, and I was only maybe fifteen at the time. And from that moment on his childhood was marked by being in and out of the system."

Not that Hazim was doing well either. By the end of his junior year, his grades were so poor that he was in danger of having to repeat the year if he intended to graduate. Right around that time, his mother returned from Atlanta. Her disappointment was profound and Hazim was ashamed. She had risked jail to send him to a well-funded elementary school. This was how he repaid her? "I felt like I was betraying her sacrifice by not living up to it," he says today. "And from that moment on I sort of took off." He says he needed to make up 72 credits, essentially an entire year of school. Gwendolyn paid $5,000—a fortune for her—for him to enroll in a credit recovery program, which took place during the summer before his senior year and into the fall after school.

After a teacher gave him a book called *Tyrell* by Coe Booth, the story of a young Black boy growing up in a homeless shelter and trying to help his mother make ends meet, reading became a lifeline. He had never connected with a book before like he did this one, seeing himself in the story. He began to enrich his world with words and came across a particularly meaningful James Baldwin quote: "You think your pain and your heartbreak are unprecedented in the history of the world, but then you read."[32] By senior year, at a specialized charter school he attended for just one year, Hazim was excelling academically. "At that point, I was hungry," he says, and he fell in love with learning, "not just because of the content, but this feeling of being able to enact self-transformation, through learning. I saw that happen in my life, in real time." He stopped hanging out at the basketball court and zeroed in on school.

Although he had done markedly better senior year, Hazim's overall GPA was weighed down by his early years and still so low (2.3) that he couldn't go straight to college. Instead, he enrolled in community college, where he was initially placed in remedial classes but soon switched into the honors program. He began to sit in the front row during class and meet with his teachers. Like his elementary school, Community College of Philadelphia was a place where students didn't need to prove they were smart enough to be there. They were given space to be creative with their projects and analysis. Hazim read voraciously. At the library, if he found a book that interested him, he'd go on to read all the books shelved in the same section. His bedroom windowsill was soon stacked with books— philosophy, pedagogy, fiction, African American studies. It was all grist for his new learning mill. His teachers noticed. "Ever see a kid eat and see him pack all that away and wonder where it's going? Hazim was the same way with information," one of his CCP advisors told *The Philadelphia Inquirer*.[33] He began to distinguish himself in other ways as well, becoming student body vice president and appearing as one of three outstanding first-generation college students on the Philadelphia-based national public radio show *Radio Times*.

After two years, he was ready for Temple University in Northeast Philadelphia. Just a few blocks from his childhood home, Temple had always felt both part of the neighborhood and unattainable. The school had been founded in the 1880s by a pastor named Russell Conwell. After tutoring working men and women in the evenings, Conwell saw a need for a school focused on providing education to students regardless of their background and means.[34] Conwell himself was famous for a speech he gave regularly over the years that recounted an allegorical tale, the Acres of Diamonds, about a man who is lured away from a contented life by a search for glittering diamonds when it turns out there were diamonds on his own land all along. All too often, the moral of the story goes, we overlook the human potential in our backyards. True to the parable, Hazim quickly became a star at Temple. Everyone knew him on campus. He impressed his professors in the honors program and attracted his peers with his intellect and wit. He was a manifestation of Temple's mission, the overlooked diamond finally allowed to shine.

And now we return to the beginning of the story. What does Hazim have in common with all those illustrious people I named earlier? The Rhodes Scholarship. He and those boldface names all won what is widely considered the most prestigious academic scholarship in the world. In 2018, Hazim was one of thirty-two American Rhodes Scholars.[35]

A Rhodes Scholarship provides grantees with full funding to study at Oxford University for two to three years. Most scholars come from elite backgrounds, schools like Harvard and Stanford. Hazim was the first community college graduate ever awarded the scholarship, as well as the first from Temple. When he heard his name called as a winner, he exulted, but he also felt a sadness that surprised him, brought on by his "awareness that far too many people with similar backgrounds as me remain locked out of such opportunities," he said. They simply don't have what they need to get ahead. "I don't think there's anything wrong with my community except the foot on its neck," he once told a reporter (using a phrase of Malcolm X's that later came to be associated with the death of George Floyd in May 2020). "I don't think my community needs a savior. I think they need resources."[36] In a later conversation with me, Hazim highlighted the often-overlooked richness of his community and the fact that stimulating language environments come in different forms. It was a neighborhood much like his that inspired the rapper Jay-Z while he was growing up in Brooklyn. "It was the soundscape of the Marcy Projects that gave birth to his 'verbal imagination,'" Hazim said.

After two years at Oxford, he started a PhD at Harvard in American Studies, one of only thirty-nine doctoral students in the department. He seems poised to produce revolutionary scholarship in Black Studies, to add to the work of his literary heroes, such as bell hooks, W. E. B. Du Bois, and James Baldwin. In North Philadelphia, his old neighbors certainly think he is destined for great things. As he walked the streets recently, a childhood friend called out, "Hazim for president!"

Now, you may think Hazim Hardeman's story is so unusual as to be beside the point. After all, there is no list of lost Einsteins to point to—no Google search will turn them up, because they are by definition lost and Hazim is the exception who was found. As a scientist, I appreciate that a story like Hazim's—as inspiring as it is—lacks generalizable data. So let me share some actual data, really a single graph, that captures the stories of babies born in the United Kingdom during a single week in 1970. It puts into stark relief the likelihood that lost Einsteins will be . . . well, lost.

Average rank of cognitive test scores at 22, 42, 60, and 120 months, by socioeconomic status (SES) of parents and early rank position

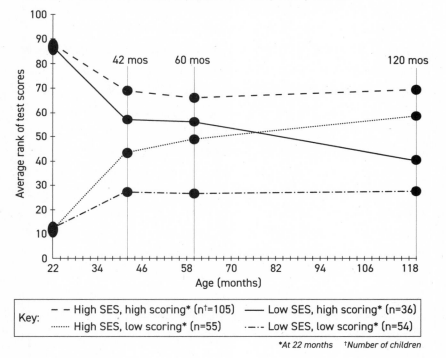

Key:
- - High SES, high scoring* (n†=105) — Low SES, high scoring* (n=36)
······· High SES, low scoring* (n=55) · — ·· Low SES, low scoring* (n=54)

*At 22 months †Number of children

This graph, which relates socioeconomic status (SES) to changes in cognitive test score rankings over a period of about eight years, is just a small piece of evidence compiled as part of a much larger project. Beginning in 1946 and repeated in 1958, 1970, 1989, 2000, and 2020, the British have conducted longitudinal cohort studies of thousands of children born at the same time.[37] These studies have significantly added to what we know about a whole host of societal concerns such as education, obesity, and mental health. The graph above comes from a study of a subset of the seventeen thousand children born in 1970 who are being tracked by researchers.[38] It speaks to exactly the question we're considering—the impact of socioeconomic status on children's intellectual ability. The first measure relied on cognitive test scores from just before the age of two until the children turned ten. The researchers used the children at the high and low ends of the distribution of intellectual quartile (a measure of how children compare to the rest of the group). Early on, there is a healthy mix of rich and poor at both the top and the bottom. By the age of ten, however, a major shift has occurred. Children from higher-income families who start at the bottom of the intellectual scale rise in the test-score rankings—easily overtaking the gifted low-income children and closing in on their higher-income peers who had started at the top of the intellectual pyramid. Meanwhile, an opposite fate befalls the children from the poorer families. Those who start out at the low end of the scale stay there. And the children from lower-income families who start at the top cognitively? Well, they drop dramatically to end up scoring not much better than their less-gifted peers. These are the potential lost Einsteins. This captures a truth a *Washington Post* reporter summed up this way: "It's better to be born rich than gifted."[39]

Such findings make Hazim Hardeman all the more remarkable. Through love and determination, it seems his mother built the foundation for his remarkable brain. But what if she had played by the rules and hadn't allowed him to see a vision of what was possible? What if he hadn't "turned left when his friend turned right" and had landed in jail? What if Gwendolyn hadn't been able to pay for the credit recovery program? It could so easily have gone another way for Hazim. In other children's lives, those "what if's" become the thousands of datapoints coalescing into the devastating truth portrayed in that graph.

Hazim recognizes that many people who hear his story will focus on his intellectual triumph against the odds. His is, after all, the quintessential Hollywood feel-good story. When I think of Hazim, though, I don't see a feel-good story—even though it is! Rather, just as Hazim himself felt conflicted when he learned he had won a Rhodes Scholarship, I grieve for the millions of kids—the lost Einsteins and also the Average and Above Average Joes and Janes—who have been denied the opportunity to fulfill their fundamental promise, however great or modest it may be. I see the myriad moments when this Rhodes Scholar almost wasn't. Hazim has a powerful message for anyone who misses the larger import of his story. "Don't be happy for me that I overcame these barriers," he says. "Be mad as hell that they exist in the first place."

The North Star

Parents want the same thing: to provide their children every opportunity in the world. That's what I wanted when I saw myself standing at the riverbank in my dream, to clear a path to whatever fulfillment my children's ability and hard work could afford them. I see the same primal desire in Hazim's mother and Charlotte's parents and I imagine their fears as they considered the challenges facing their children, as they worried about what lay ahead. It is this desire that unifies us as parents, whoever we are, rich or poor, Black, white, or Brown, living with disabilities or not.

The brain is primed to do what every parent wants—to prepare a child to take advantage of all the opportunity that its neuroplasticity offers. Providing opportunity is tied to that neuroplasticity during the first three years of life, when it is at its highest. You may well ask why, then, if the brain has such incredible capability, do things go awry for so many children, as we saw in Kim Noble's work and the British Cohort study? It isn't the brain that is getting it wrong, it is our society. Those three pounds of precious neural tissue come with limited instructions. For much of what it will do, the brain requires guidance from the world. And herein lies the problem and the root of the invisible epidemic: The brain builds for the world it is born to. It adapts. If the world that a brain is born into lacks resources, the brain lacks resources, and will suffer accordingly.

What we must not do is blame either the parents or the babies. "People often say, it's not poverty, it's whether you're willing to pull yourself up by your bootstraps," Kim Noble says. "Well, babies don't have bootstraps."

What babies do have is adults—their parents, their communities, and society at large. Together, those adults are capable of pulling those babies up. Together, they can make healthy brain development society's new North Star. Together, they can ensure that each child has what is necessary to fulfill the promise of their promise.

THREE

THE STREETLIGHT EFFECT

"Your whole life, sir. You have followed the wrong star."

—BLIND MAN CASSIDY[1]

Maybe you've heard this one: Late one night a police officer happens upon a man crawling around on his hands and knees under a streetlight.

"Sir, what are you doing?!" the officer asks.

"Lookin' for my keys," the man answers.

"Where did you drop 'em?" the officer asks as he bends down to help.

"Across the street."

"Then why are you looking over here?!"

"Light's better."

Variations on this joke show up frequently in science. The reason it resonates so strongly is that nearly all scientists have had the experience of seeking answers in the wrong place, choosing to look somewhere because, frankly, it's easier than the alternatives. There is an almost unconscious pull toward the light. The technical term is "observational bias."[2]

We measure what is most measurable. We work with the population that is most accessible (often college students). We treat the most obvious symptoms. We look where everyone else is looking. Sometimes that bright spot is completely irrelevant to the question at hand, as with the man who'd lost his keys. And sometimes the light provides only partial, and even misleading, answers. The stories abound. In Alzheimer's disease, researchers have spent decades designing drugs to tackle the plaques they were convinced were the primary problem, in part because they could see them. They've had depressingly little success and only recently have gotten serious about other avenues of attack.[3] In business, people exert endless time and energy on generating likes and follows on social media—again something they can see, and measure—even though such viral popularity hasn't been shown to lead to a tangible increase in sales.[4] The result of this "streetlight effect," as one writer put it, is that we "tend to look for answers where the looking is good, rather than where the answers are likely to be hiding."[5]

The implications for the future of millions of young children are serious. In many countries around the world, the problem people have been trying to solve—the key we have been looking for—is this: How do we raise the next generation of productive citizens? How do we ensure that all children have a fair and equal chance at realizing their full potential? The usual answer is education, meaning school. "Throughout the globe, education is now recognized as the new game-changer that drives economic growth and social change," former U.S. Education Secretary Arne Duncan said in 2011.[6] The specifics of how to deliver that education vary from country to country—more days of school in South Korea, aggressive teacher recruitment in Finland and Singapore—but, ultimately, they are variations on the same theme. In other words, we consistently look for answers in the same place: K–12 education.

Here in the United States, the goal of giving children a fair shot in life is the beating heart of the American Dream, which declares ours a country where each and every one of us can succeed regardless of the circumstances of our birth. It says we are a meritocracy, not an aristocracy. In truth, many other countries espouse the same idea even if they don't sum it up so neatly in a national catchphrase.

But the American Dream is fading. All that opportunity was supposed to lead to generational mobility. It held that most of us would do better than our parents—that we would be better educated, earn more money, move up in the occupational ranks. But the fraction of children who earn more than their parents has fallen from approximately 90 percent of children born in 1940 to 50 percent of children born in 1980, according to Harvard economist Raj Chetty.[7] Overall occupational mobility has declined by similar numbers. Of course, not everyone can or should be a surgeon or a lawyer, but it used to be that occupations that didn't require a college degree could comfortably provide for a family. Most of those blue-collar occupations don't pay enough to do that today.

This is not what we were promised.

For as long as we have been a nation, we have thought that education, by which I mean formal schooling from the age of five or six up to eighteen—K–12 in today's terms—was the way to achieve that dream, to deliver the social mobility we thought was our birthright. Over the centuries, public education has been held up as a means of building a cohesive nation, producing informed citizens, assimilating immigrants, and providing opportunity to all. For much of our history, of course, "education for all" did not literally mean "all"—girls, Black people, and Indigenous people were excluded. But the idea was celebrated. The way to avoid tyranny of the powerful, wrote Thomas Jefferson, was for the government to provide basic education for all in order to "illuminate . . . the minds of the people at large."[8]

We have thrown billions of dollars at the task of delivering quality

K–12 education and worked through a raft of reforms. Yet in recent international tests of reading and math like the Program for International Student Assessment (PISA), American students continued to hover halfway down the list of developed countries. Our ranking reflects essentially stagnant progress over the last twenty years. Some students improved (strong readers did better), but others did worse (the bottom tenth percentile scored lower). In the 2019 National Assessment of Educational Progress, only one-third of American children were proficient readers.[9] Two out of three kids can't read proficiently! "It just isn't working," education professor Daniel Koretz, of the Harvard Graduate School of Education, told *The New York Times* late in 2019, of the overall effort to improve our standing.[10] Even so, because we are so sure that K–12 education is the great equalizer, because K–12 is what sits under the streetlight, we keep investing heavily in it, and not much else.

What gets lost in the shadows on the dark side of the street? What "key" truth are we blinding ourselves to? Education begins on the first day of life and not the first day of school. By the first day of kindergarten, there are real differences among children. Some arrive ready to learn. For others, it's already remedial. Time and again, we have failed to look beyond the glare of the light to appreciate the importance of the first years of life and the necessity of getting those years right.

Why We Don't Begin at the Beginning

Modern education's focus on K–12 has an origin story, one that goes back a surprisingly long way. John Amos Comenius, a seventeenth-century minister born in what is now the Czech Republic, laid the groundwork for the system we have today.[11] Born in 1592, Comenius, like everyone else of his time, thought education should be rooted in religion—learning to read was necessary in order to read the Bible. But some of his other ideas about education sound so modern today, they must have been truly radical in the 1600s. He argued for learning in students' native languages, and looked to meet children where they were, developmentally, by creating the first textbooks with pictures and proposing that teaching should progress from simple to more complex concepts. He urged teachers to pay attention to the minds of individual children and the way they learned, "following in the footsteps of nature."[12] Equally radical for his time, he called for education for rich and poor alike, as well as for women. Comenius also advocated for three tiers of formal education that looked a lot like today's division of elementary school and high school followed by university. His ideas, even before the publication of his famous 1657 treatise on education, *The Great Didactic*, went the seventeenth-century equivalent of viral, quickly spreading across Europe and, thanks to the Puritans who carried them there, to the American colonies. In short, Comenius was an education "influencer" of his day.

In 1642, the Massachusetts Bay Colony passed a law requiring parents to teach their children to read. Then in 1647, they passed a second law that would be replicated by other colonies and laid the ideological groundwork for public education in America. It was called the Old Deluder Satan Act—yes, you read that right. The theatrical name came from the Puritans' belief that the devil wanted to keep people illiterate. To create a productive society, they believed, each community member should be able to read and understand the Bible, as well as local laws and ordinances. Satan, that old deluder? His evil will could be countered through mandated universal education. The act required any community of fifty households to hire a teacher and any town of one hundred families to build a "Grammar-School" for their children, though in practice these grammar schools were for boys only, beginning around the age of seven. Costs were to be shouldered by parents or by the larger communities.[13] The system largely mirrored Comenius's recommendations.

But for all his innovation, Comenius overlooked one crucial ingredient: the first years of life. Even then, the streetlight (or, rather, the oil lamp) was focused on older children. Comenius didn't just overlook early childhood education, he argued against it on the grounds that it could be harmful. "Then it is safer that the brain be rightly consolidated before it begin to sustain labors; in an infant the whole cranium is scarcely closed, and the brain is not consolidated before the fifth or sixth year," he wrote.[14] That idea took hold and has endured.

It is understandable why Comenius and those who have followed thought that school should start at age six or seven. Developmentally, there is an important change in cognitive ability and maturity that occurs around that age. The change is widely recognized—by developmental psychologists, anthropologists, evolutionary biologists, and others.[15] Known as the "five-to-seven shift," it marks the growing sophistication of what children can do cognitively as they become less exclusively concrete in their thinking and more capable of logical reasoning. The change is so obvious you don't have to be a scientist to see it; you just have to be a parent. In hunter-gatherer societies, somewhere between ages five and seven brings the introduction of increased responsibilities for children. Of course, focusing exclusively on the shift that occurs in the five-to-seven years overlooks the brain-building that has to be done in the years leading up to that point in order to accomplish the shift.

Not that Comenius believed that children couldn't or didn't need to learn anything before they turned six. To the contrary, he thought the role of mothers (he focused on mothers, not fathers) was critical. But he believed that the learning that had to be done in the early years was a private affair. So, when Comenius wrote his *Great Didactic* and outlined a modern system of public education, he argued that it is a mother's responsibility to educate children under six and

the state's responsibility to educate children over six. And throughout history since then, parents have had to figure out how to make
that learning happen during those formative early years.

One of my favorite examples of this is the Sewall family of
Puritan-era Massachusetts. Their story reminds me just how enduring parental struggle and worry have been. Mothers and fathers have
always wanted to give their children the best start in life. Samuel
Sewall was no different. He was an English immigrant who arrived
in the American colonies at the age of nine. He started at Harvard
in 1657, where he almost certainly read Comenius's *The Great Didactic*, a required part of the curriculum. Sewall married into a wealthy
family and became a memoirist and judge. He advocated for the
abolition of slavery and presided over the Salem witch trials (for
which he later expressed profound regret). He and his wife, Hannah, had fourteen children. So enamored was Samuel of his children's development that he wrote in his diary about the love and
admiration he felt when his eighteen-month-old son, Hull, said his
first word: "apple." Alas, Hull was sick for most of his short life and
died just six months after his father's diary entry. Like Hull, six of
the Sewalls' other children died before reaching adulthood and one
was stillborn—a tragedy that wasn't unusual for the time.[16]

As you might imagine, the deaths of so many of their children
caused the Sewalls great suffering. It may also have intensified their
dedication to giving their surviving children the best chance possible, which they did by sending them to the local grammar schools
established by the Old Deluder Satan Act. Samuel Senior had high
hopes that his namesake and oldest surviving son, Samuel Junior,
would follow his father to Harvard and then perhaps go into the
ministry. But young Sam was not suited for Harvard. By most accounts, the boy started grammar school at eight and by sixteen—the
age at which his father was a Harvard grad twice over—the younger
Sam was skipping school.

I picture Hannah and Samuel Senior tossing and turning at night.

Perhaps one leans over to their bedside table, lights the lamp, and illuminates both their faces. In hushed tones, they agonize over what to do with Sam. Then, after some discussion, they settle on a plan: They'll pull him out of school and send him to be a bookseller's apprentice. Maybe it didn't happen that way, but I do know the universal love and worry that all parents have for their children—even the Puritans, who recent scholars say we've unjustly stereotyped as being especially harsh.[17] (One of many unjust judgments made of parents in my experience!) What we know happened was that the Sewalls arranged for their oldest to apprentice with a bookseller, Michael Perry. But bookselling didn't take, either. Samuel left the apprenticeship after a few months and spent a wayward decade before settling down as a farmer.

Were Hannah and Samuel relieved when young Sam finally found his way? I hope so. Did they wonder what they could have done differently to give their son a stronger start in life? Probably. I do know that Samuel Sewall Sr. changed up his approach with his next surviving son, Joseph. In particular, he started Joseph's education earlier . . . much earlier. In this, Sewall was following the lead of his good friend Cotton Mather, a well-known minister and author. Sewall knew that while Comenius criticized early schooling, Mather strongly disagreed. "No. You can't begin with them too soon," Mather said. "*Quo semel est imbuta recens servabit odorem, testa diu.*"[18] That is Latin for "The first scent you pour in a jar lasts for years."

So, at the tender age of two, the spirited Joseph, born a decade after his brother Sam, was sent to a "dame school." Officially, these schools focused on what they called the "four R's": 'Riting, Reading, 'Rithmetic, and Religion. But in reality, many of the dames did little teaching. For their efforts, the dames earned meager wages. The town of Woburn, for example, offered one local teacher, Mrs. Walker, ten shillings for a year of teaching but then deducted seven shillings for taxes and more for produce and other expenses.[19] That left poor Mrs. Walker with all of one shilling and three pence for her work, though she may also have bartered with some parents (how does two pieces of firewood for two weeks of instruction sound?).

The quality of these schools varied enormously, but they got the job done for Joseph, who cycled through a series of dame schools (apparently the better ones) and also studied for a while with a notable reverend of the time. The Sewalls wanted to keep Joseph "from falling into the undisciplined habits that had hobbled Sam in his studies," according to one historian's account.[20] At ten, Joseph started at the prestigious Boston Latin School, where boys of a certain class went to prepare for Harvard. By age eighteen, he had graduated from Harvard (valedictorian!) and, three years later, obtained a master's degree. Joseph became the minister his parents had been dreaming of. He enjoyed the work so much that he later turned down the presidency of Harvard to continue in it.

At first blush, it appears the moral of the Sewalls' natural experiment is that you need to start early (true) and that if you do, your child will go to Harvard and have the career of your dreams (no guarantees on that part). But I see something more telling. In the haphazard dame schools, we can see the very beginnings of the system that has left our youngest children—their education and their brain development (which, to be fair, no one properly understood at the time)—in the hands of women who had not been trained for the job. To this day, it is the case that poorly paid, largely untrained, and often socially marginalized women are doing the job of caring for and educating many of society's youngest children, including the children of many middle-class and affluent parents. Despite ample evidence showing us how critical their labor is, they receive neither the respect nor the pay commensurate with the importance of what we ask them to do.[21] Basically, we pay them babysitter money. And for those who can't afford even that, some patchwork of a support system—grandparents, older siblings, other family members, neighbors—must step in to take care of the little ones while their parents work the multiple jobs required to pay the bills.

The More Things Change . . .

When it comes to parents and early childhood, the more things change, the more they stay the same. I first met Mariah when she came to TMW with her son Liam, who was one year old at the time. She already had a great way with kids and added the ideas we shared in our sessions with her in no time. Watching her play with Liam and talk with him, I could see her energy and creativity. He climbed into his mom's lap to join her as she picked up a book about a baby lion called Chomp. "The lion *ROARS*," Mariah read in her booming voice, which could easily fill an auditorium. Liam plucked the book away from her, and Mariah, sensitive to her young child's need to feel he was in charge, let Liam take the lead. He flipped through the pages one by one, no easy task with such tiny fingers. Meanwhile, Mariah picked up another book, *Ten Little Fingers and Ten Little Toes*, and read it to him.

"There was one little baby that was born in the hills, and another who suffered from sneezes and chills," she said in a clear, singsong voice. "And both of these babies, as everyone knows, had ten little fingers and ten little toes."

Mariah grabbed Liam's fingers and then his toes, tickling him to make the connection between the words on the page and his own body. Liam squealed with delight. Together, Mariah and Liam turned their attention back to Chomp. Still wanting to be in charge, the little boy flipped through the pages till he reached the last one.

"Good job!" Mariah said as Liam reached the end.

Liam babbled an incoherent string of words in reply, but one was clear. "Job!" he said, his tone showing how proud he was of his accomplishment.

For as long as she could remember, Mariah wanted to work with young kids. Born and raised between Chicago's South Side and its suburbs, she grew up in a large extended family—her grandparents were married for sixty years and had thirteen children, who had multiple children of their own. Growing up, everyone was exceedingly close. Mariah called her cousins her brothers and sisters, and she loved her big family. It meant she had plenty of playmates, and later, when the older cousins and siblings started having children of their own, a good deal of babysitting practice. "I always loved children," Mariah said. "I've been raising kids all my life." She couldn't wait to have her own children and assumed she knew what parenting would be like.

But her own kids, Liam and his little brother, Lain, proved surprisingly challenging. "It confused me so much when I had my own because I just thought that I had it all figured out," she says. "I have been raising kids forever. I've been watching kids. I got this. No. [It was] different." That recognition, and a desire to learn and grow as a mother, was what drew her to TMW's home visiting program. "I wanted to get more depth in knowing what to do with my son," she says.

As it does for so many parents, new motherhood had left Mariah overwhelmed and exhausted. But the main source of her emotional and physical depletion was her job. When her boys were both still under two, Mariah started working at the center where she sent them for childcare. Her relationship with their father hadn't worked out, so she found herself having to support the family alone. There was a lot she loved about the new job. Early on, she was placed with the older preschool children, and she noticed a young boy I will call Joe sitting in the corner, playing by himself, not interacting with any of the other children. Joe was autistic and nonverbal, but Mariah went and sat down near him and kept up a lively monologue, as if Joe could have a conversation with her. "I treated him like he was my own," she says. Her effort paid off right away. "He would make little noises and he would hug me, and he would try to kiss me." Mariah found she had a special knack for working with children like Joe. "I noticed those children gravitating more towards me." As an assistant teacher at the center, she spent much of her time caring for the children with special needs. The work felt like her calling. "You just have to have a special type of heart for those kinds of kids," she says.

But passion couldn't pay her bills. Mariah made only minimum wage, and the job offered no insurance or benefits. Living in Chicago, as a single mother of two young boys, her salary was the contemporary equivalent of trading firewood for education. "I was literally working and taking care of my babies. Nothing else. My checks went to bills, bills, bills, bills"—for public transportation (she had no car), rent, groceries. "I never really had money to do anything outside of that," she says. Even living frugally, she never had enough. Every day was a struggle. Mariah argued with herself regularly about getting a better-paying job. The internal debate went something like this:

This is not enough money.

But I love my job and I love those kids so much.

Sometimes she imagined different paths, the ones she had almost taken. She went to college for a few years, where she studied criminal justice, but she'd decided that wasn't for her. Inspired by a cousin, she considered being a mortician. The cousin earned $900 for each body she embalmed. Caring for babies, Mariah would have to work for almost three weeks to make what a mortician earned in an afternoon. (Something about that seems terribly wrong—that society would pay so much to embalm the dead, while paying so little to support the living.) But when Mariah went to her cousin's workplace to get an up-close look at what she did and saw her embalming a teenage boy who'd been shot in the head, she felt the tragedy of it, the loss of potential. She knew then that her commitment had to be toward building a future for the children in her care. That's what she loved.

But the old adage "If you love what you do, you'll never work a day in your life" doesn't hold up when what you love doesn't pay enough to feed your children dinner. "I wasn't living, I was just existing," Mariah says of that time. "It was like I was a robot. I'm getting up, I'm doing this, and I'm doing that. I'm getting maybe four hours of sleep and I'm doing it again." Crying babies filled her days at work and her nights at home. She got sick all the time, and she went to work sick because she couldn't afford to take a day off. Giving up felt like letting her students down, letting her babies down. "I never believed in taking time for myself," Mariah says. "I was losing weight. I was stressed out. I was lifeless." She lost touch with lots of friends, always turning down their invitations. Even her friends from work, who were paid just as poorly, didn't have kids and so didn't understand the double burden. Most of the time, Mariah felt bad for feeling bad. "I should be happy. I should be enjoying [my babies]. I should be enjoying life." But the push and pull of the constant demands in her life left her feeling, on the one hand, like a bad parent if she ever went out or was away from her kids and, on the other hand, like a bad teacher if she contemplated quitting her job.

"It Just Isn't Working"

In the centuries between Samuel Sewall's time and Mariah's, blinded by the streetlight, we have stayed true to our tunnel-vision focus on K–12 education as the way to give every child a chance, while ignoring the needs of our youngest children, whose brains are in their most formative period. Occasionally, we have been jolted by the realization that it isn't working very well. In 1957, when the Soviet Union launched the space age by putting the Sputnik satellite into orbit, Americans were shocked to realize we were not the unquestioned international leaders we thought we were. "Suddenly they were in outer space before we were. And how could this have happened?" said educator Chester Finn in *School: The Story of American Public Education.*[22] The most obvious answer to everyone was that the Soviets were better educated. What followed was an aggressive push to improve math and science in American schools via the National Defense Education Act. (The Soviet Union also had extensive early childhood education, beginning at age three, though that didn't seem to register with those concerned about why we had fallen behind.)[23]

Fast-forward twenty-five years to 1983. We were jolted anew by a report that delivered another dramatic indictment of American schools. Spurred by concerns about the competitiveness of the nation's workforce, the secretary of education had commissioned a task force to assess the state of public education. The commission's report, *A Nation at Risk*, told a bleak story of failing schools that were falling behind on the world stage. "Our once unchallenged preeminence in commerce, industry, science and technological innovation is being overtaken by competitors throughout the world," the commission wrote. "If an unfriendly foreign power had attempted to impose on America the mediocre educational performance that exists today, we might well have viewed it as an act of war." Retrospectively, some believe the report wasn't entirely fair. But it had a clear result—yet another effort to improve K–12 education.[24]

In the last few decades, the United States has spent billions on one major educational reform strategy after another—No Child Left Behind, Common Core State Standards, Race to the Top, Every Student Succeeds Act, and more. We have increased federal funding.

We have introduced new forms of testing and accountability metrics for teachers. We have expanded after-school programs and decreased classroom sizes. We have pushed to increase teacher quality. We have established charter schools and vouchers. We have changed the way the school day is scheduled. We have pushed to raise standards.

As we began to realize that all these efforts still weren't moving the needle, we even expanded the reach of the streetlight by pushing for pre-K for all in hopes that eventually every child will begin school no later than four. Universal pre-K essentially adds an extra grade to the public education system and, as of 2021, pre-K for all or something very close to it exists in nine states and several cities, including San Antonio, New York City, and Portland, Oregon. Will this be the magic bullet we all hope for? I agree that pre-K is critically important. And since the campaign enjoys bipartisan popularity, it's certainly politically expedient to focus on getting a win on a year of pre-K.

But let us not fool ourselves. Pre-K for four-year-olds does not address the needs of very young children and their families, nor does it fix the tragic disparities in brain development that show up before the age of three, as we saw in Kim Noble's work.

Education Begins on the First Day of Life

To close the gap in these early disparities, and to strengthen early brain development in all kids, we must now focus on what has to happen during the first three years of life. That phase must be viewed as part of the educational continuum.

The only attention that critical period usually gets is in the context of efforts to alleviate the effects of poverty, and to address major societal crises. The earliest organized childcare dates to the late nineteenth century, when charity-run "day nurseries" were created to care for children of low-income families while their parents worked or sought work.[25] During the Great Depression, the United States government stepped in for the first time to fund childcare centers in order to put parents back to work (because of the far-reaching economic devastation, a broad subset of families was eligible). And then when World War II forced women into the workforce to replace the men who had gone to fight, the government again set up an extensive network of childcare centers, most of them remarkably high quality, as part of the Lanham Act. Mothers loved the centers and wanted them to continue, but when each crisis ended, so did the funding for care.[26] A handful of people did recognize the mistake in that. Eleanor Roosevelt, an early champion of working mothers and modern families, wrote of the childcare centers that closed after WWII, "A few of us had an inkling that perhaps they were a need which was constantly with us but one that we had neglected to face in the past."[27] But most politicians of the day made clear how much they disapproved of the government's foray into childcare: "The worst mother is better than the best institution," New York City mayor Fiorello La Guardia famously said in 1943.[28]

By the 1960s, however, knowledge about child development was growing almost as fast as the babies being studied. The importance of high-quality parenting showed clearly in the research on attachment theory by British psychiatrist John Bowlby, and in a series of pioneering studies that followed families for years. Then a handful of educational psychologists published groundbreaking work demonstrating that intelligence could and did change with experience. Suddenly, it was undeniable that environment and experience mattered from the first days of life.[29]

Born of both this new understanding and President Lyndon Johnson's War on Poverty, the groundbreaking Head Start program was launched in 1965 to help prepare the poorest children for school. (*Sesame Street* was created with the same motivations in 1969.) Although Head Start has been the subject of intense political debates over the years and there have been mixed reports on its effectiveness, it has had staying power—a hopeful sign of society's unwillingness to abandon the most disadvantaged children.[30] (That said, in 2018, according to the Children's Defense Fund, Head Start served only half of eligible three- and four-year-olds, and the Early Start program, a later addition for infants and toddlers, reached a startlingly low 8 percent of eligible babies.)[31]

While Head Start was aimed at the children of the poor, the extensive press coverage it received, and the new discoveries being made about child development, sparked an interest in early childhood programs for all children. Middle-class parents figured that if cognitive stimulation was good for poor children, wouldn't it be good for their children, too?[32] However, there were no funds available to support any kind of public system of early education and childcare. Instead, a hodgepodge of private childcare options—some high quality, some little more than glorified babysitting—came into existence for those who could afford it.

Meanwhile, more and more women were moving into the workforce. Between 1950 and 2000, women's labor force participation rate nearly doubled, rocketing from 34 to 60 percent. By 2020, it was at more than 70 percent.[33] This shift forced further debate over the question of who was going to care for young children. The inaction on childcare in the United States clashed with the realities of working mothers (and fathers), who were not going to be able to do everything that was required to help raise young children while holding down their jobs. They needed high-quality care and education programs to complement their own love and nurturing.

That imperative has only deepened as our knowledge of early brain development has continued to expand. With each subsequent decade, we have learned more and more about the critical importance of stimulating the brain during the earliest years. By the 1990s, proclaimed the Decade of the Brain by Congress and the president because of the explosion in neuroscientific knowledge, the idea that early childhood mattered had permeated the societal groundwater. It was now understood that children who are ready for school are ready for life. Add to that economist James Heckman's work in the 2000s showing the return on investment that quality early childhood programs reap for society, and there was an ample pile of evidence making it clear that quality programs didn't only benefit children but paid off for society as a whole.

Other countries are managing this critical job of school readiness far better than the United States. In 2020, the Organisation for Economic Co-operation and Development, which runs the international testing program in reading and math for fifteen-year-olds, published a pilot study for younger children. It tested five-year-olds for school readiness in the United States, England, and Estonia. The American kids came last in both emergent reading skills and emergent numeracy.[34]

After reading about this early learning study, I was curious about how some of the countries that regularly top the international testing charts handle early childhood. Finland is a striking example. Once quite poor performing, with low-quality schools and large inequalities among students, it has turned around its education system to such an extent that it is the envy of many other countries and the subject of much academic analysis and glowing journalistic accounts of its success story.

I was especially eager to know what Finnish children do before they start grammar school around the age of seven. To begin at the beginning, Finnish parents have generous paid parental leave that covers much of the first year of a child's life. Then when babies are about nine or ten months old, parents can choose between public or private Early Childhood Education and Care (ECEC) programs, or they can choose to stay home with their kids, and if they do, they get a stipend that helps support that choice until the youngest child is three. At least one year of an ECEC program is mandatory to make sure all kids are school-ready.[35]

If I had to guess, I'd say that the secret to the Finnish success— one piece of it anyway—is hiding in plain sight. It's right there in the name of their early childhood program: Early Childhood *Education* and *Care*. In much of the world and certainly in the United States, there is a long-standing sense that care is babysitting and education is school and never the twain shall meet. In other words, there is a hard distinction between formal and informal care and education. But the Finns do not distinguish between "care" and "education." In fact, the nickname for their integrated program is "educare," which has also been adopted by groups like the early childhood organization Start Early in the United States, and hits the nail on the head as to what is developmentally appropriate.

Think about it. When my children were toddlers, and I took them out in their strollers for a walk, I routinely pointed out the birds I spotted, the buses that rolled past, and the big yellow dog our neighbor was walking. I'm sure most parents do something similar—at least if they have the time. Is that care or education? It is both.

Looking beyond the Streetlight

If the Sewall family could engage in time travel, what would they make of "educare"? I think they would have trusted that any child who got to enjoy it would be well prepared for grammar school and would have been thrilled to send their children to such a program.

As for Mariah, when I described the Finnish system to her, she said emphatically: "We should move to Finland." Fortunately, she didn't have to, because she found a better childcare job for herself. After Mariah spent a year struggling to balance the effort of supporting her boys on her own and working at that first childcare center, her mother saw how miserable her daughter was and how hard she was working just to stay above water financially. So she bought a house in the suburbs and encouraged Mariah to move in with her boys. In the new house, Mariah had the space to breathe, reset, enjoy her babies, and focus on finding a better-paid job in a field she loved.

She discovered a childcare center that had recently opened in the new neighborhood. It was started by a former police officer named Felicia who had grown bored of retirement. Felicia hired Mariah as an aide and Mariah loved the new job. Felicia was organized, held her employees to a high set of standards, paid them well, and offered benefits including good insurance. Demanding and practical, she also nurtured her employees and helped them figure out how to advance their careers. As Mariah tells it, Felicia would say: "If you don't want to be an aide, 'Okay, well [what do] you want to be? A teacher? I'm going to show you what you need to do. You want to go up on pay? I'm going to show you what you need to do to get up on pay.'" The new job was a radically different experience for Mariah and one she valued. "I have a boss that sees me, and she sees my potential," she says.

The way Felicia runs her business highlights what's possible. But given how few Felicias there are, we can't count on them, and society must step in to help with the education of our youngest citizens. Society pays for what it values. If we value our children's futures, we should be treating childcare workers the way we treat other essential educators and public employees, not forcing them to get by on scraps the way Mariah did.

If we really want to find the key that will change our children's educational, economic, and occupational fates, we must peer into the shadows beyond the glow of the K–12 streetlight. What we see there will reveal that learning begins on the first day of life; that what happens before kindergarten matters as much as what happens after; and that if we don't focus our efforts on those first few years, we will never get where we're trying to go.

It has been hard for many to see this truth because what happens between adults and children during those first few years doesn't look like "school." It *shouldn't* look like school. Babies and toddlers don't belong at desks doing worksheets. Of course not. They aren't ready physically or cognitively for that sort of thing. But they are ready for nurturing, stimulating conversation. They are ready to interact with their world and must have constant opportunities to do that.

If our goal is truly to launch all our children into lives that allow them to grow into their fullest potential, then we must make brain development during the earliest years our new North Star. That is the real key we have been scrambling to find.

When healthy brain development is our goal, our perspective changes. Suddenly, what happens between birth and three looks like a natural part of the educational continuum. When healthy brain development is our goal, we dismantle the false demarcation between care and education, and understand that they are inseparable from each other in the pivotal years when the brain's architecture is being built. And when healthy brain development is our goal, we recognize all adults—parents, family members or friends, babysitters, and childcare providers—as brain architects during the first three years of children's lives. And as a society, we will dedicate ourselves to supporting those early brain architects.

The North Star is not the brightest in the sky. That honor belongs to Sirius. The North Star comes forty-eighth. Yet it has been the one that has guided travelers for thousands of years. Sometimes you have to look beyond the brightest light to see the right path.

FOUR

THE BRAIN ARCHITECTS

"If a community values its children, it must cherish their parents."

—JOHN BOWLBY[1]

My late husband, Don, had a very big head. Literally. I used to say that he needed that unusually large skull to contain his big brain. (He *was* brilliant). And we had a running joke that any son of his would no doubt inherit that sizable noggin. So, when I was heavily pregnant with our son, Asher, and our obstetrician told us it looked like I was going to have to have a Caesarean section, Don and I looked at each other and burst out laughing. *We knew it! His head wouldn't fit through the birth canal!*

Ironically, the size of Asher's head turned out not to be the problem. He was breech. Stubborn even then, Asher was turned the wrong way in my womb and refusing to move. I had the C-section, and all went well.

Even though Asher's head wasn't the issue in my pregnancy, head size *is* a significant factor in the evolutionary story of human babies. A horse or zebra foal can stand up within minutes of birth and walk within two hours. Baby chimpanzees can cling to their mom as she jumps from tree limb to tree limb. And, on day one, newly hatched sea turtles know how to find their way from the sand where they're born into the sea. But human infants arrive in the world a bit under-done. So much so that pediatricians call the first three months of a baby's life the "fourth trimester." Newborns can't hold up their heads for months. Walking and eating on their own takes a good year.

The leading explanation for this apparent developmental delay is that nature had to compromise. When humans began walking up-right on two legs, the pelvises of females narrowed. Forty or so weeks of pregnancy produce a baby that's at the upper limits of what a woman can pass through her birth canal, but the forty-week-old baby brain still has a lot more growing to do.[2]

Most of that brain growth will occur in the first two years of life. At one month, a baby's total brain size is only about one-third that of adults and then increases dramatically to 72 percent of an adult brain by one year, and 83 percent by age two. Most other animals are born with brains that are closer to adult size.[3] That's why those sea turtles, horses, and chimpanzees can do so much more than human infants at birth.

The fact that we enter the world with a relatively underdeveloped brain and experience this protracted period of development is one of the greatest evolutionary gifts. It offers immense opportunity for brain stimulation during the period when the brain is in its most formative stage. This period is at the heart of humans' unparalleled intelligence, creativity, and productivity.[4]

The High (Energy) Cost of Brain-Building

All that brain-building requires a remarkable amount of energy. Even for adults, thinking and brain activity take quite a bit of energy. Even resting takes energy! If you're like me, you probably don't spend much of your day at rest. What parent has time for that? But if we're lucky enough to find the time to take a break and sit quietly, our brains will *still* use up 20 to 25 percent of our body's overall energy. That translates into some 350 to 450 calories a day for women and men, respectively. Once we start working on—or thinking about—anything challenging, that energy use soars. Chess grandmasters, who are not exactly known for their physical prowess, can burn up to 6,000 calories per day just contemplating their next few moves during a tournament. (Honey, let's do a chess workout!)[5]

But nothing compares to the energy costs of a child's brain in the process of building, not even a chess grandmaster's exertions. It makes sense when one realizes that the brain must double in size in a mere two to three years. When you try to imagine the electrical activity that buzzes around a baby's brain as one million connections are formed every second—*zzzpt*, there goes a new connection . . . *zapt*, there goes another one—you can almost feel the energy being expended. A four-year-old's brain is using about 40 percent of the body's energy expenditure.[6] This is why it takes human bodies, unlike those of other mammals, so many years to reach their adult size. In children, most of the energy is going toward brain- rather than body-building.

Where, then, is all this energy coming from? While milk feeds the body, it is language and nurturing input that feed the brain. In other words, this evolutionary strategy of extended brain development is contingent on parents (and other caregivers) to bridge the gap and help babies go from helpless to brilliant.

As the saying goes, with great reward comes great risk. In the brain, what is enhanceable via this input is also vulnerable. Inherent in humans' evolutionary big bet was an understanding that someone would always be on hand to nurture our young. Because babies are so completely helpless, the potential of their developing brains is absolutely dependent on brain-strengthening input. Without it, the rate of neuronal growth slows down considerably. If connections aren't made, those neurons that aren't being used get pruned, and some opportunities are lost forever. This brain development or lack of it is an outcome over which no helpless baby has any control. Just as our prolonged helplessness is an evolutionary gift to allow us to be the most brilliant of species, the role of parents and loving care-givers is a gift, too. It allows them to be the brain architects of these helpless beings, thus giving them the opportunity to make the kind of difference in their children's lives that we all want to make.

But society seems to squander this incredible evolutionary gift.

Impossible Decisions

I know too well the heartache of children who lose a parent. My kids paid an enormous price for the loss of their father. But there are less dramatic ways—heartbreaking in their own right—that we routinely separate parents from children right when they need each other most.

Kimberly Montez knows this as well as I do. As a pediatrician, she has devoted her career to the health of children, especially those with the deck stacked against them. She was one of those kids herself. Born with a ventricular septal defect, essentially a hole in her heart, Kimberly grew up in a low-income Mexican American neighborhood outside Houston and spent her childhood in and out of hospitals and doctors' offices. Her family struggled with the crushing cost of her treatment. Her mother was an administrative assistant for the state government, and her father could not hold down a job. They survived on Medicaid, free clinics, and even a donation from a charity to cover the cost of surgery to repair the hole in her heart. Some of the pediatricians and cardiologists Kimberly met were thoughtful and caring, others not so much. She decided early on that she would be one of the good ones, a caring doctor who would dedicate herself to taking care of under-resourced people. "That was what gave me the most passion and joy," she says. With degrees from Yale, Stanford, and Harvard, Kimberly became a pediatrician in San Diego, Boston, and ultimately North Carolina, specializing in serving communities like the one she grew up in.

Within medicine, Kimberly is a leading voice arguing for the importance of paid parental and medical leave. As I've already pointed out, as of 2021 the United States is one of very few countries in the world not to mandate paid leave. But while it was Kimberly's childhood experiences that compelled her into a life of caring for others in need, it was the shock of finding herself fighting for her own child that drove her to become a social policy advocate. Her activism was ignited when Kimberly lived out every parent's nightmare.

When Kimberly and her husband, Jamie, decided they wanted to start trying for a baby, she was working at a low-income community clinic in Boston. Like a lot of female physicians (myself included), all the training Kimberly had put in meant she came to motherhood on the late side, in her mid-thirties, or as obstetricians call it, "advanced maternal age." This put Kimberly at higher risk for complications. Oscillating between elation and heartbreak, Kimberly got pregnant, then miscarried. When she got pregnant again, she and Jamie were cautiously excited but anxious. Their fears came to a head a few hours after an appointment at eighteen weeks when Kimberly had an ultrasound. Out to dinner that night with her husband, she heard her phone dinging. Her results were in. They were having a girl. "I was just bursting," she says, "I was ready to dance around the table." But her happiness soon dissipated. Her tests had turned up some possible abnormalities. More fear. More questions. More tests. All she could do was continue to breathe and wait, breathe and wait. Two weeks later, now twenty weeks pregnant, she got good news and bad news. "There's nothing wrong with the baby," the doctor told her, "the problem is with *you*."

Kimberly's heart dropped and her stomach knotted when she heard the diagnosis. Her cervix, the outer end of the uterus, was open. The condition, called a weakened or incompetent cervix, occurs in roughly 1 to 2 percent of all pregnancies and can lead to early birth or worse, a miscarriage.[7] Only halfway through her pregnancy, she was told she might deliver any day. Babies born at this point in a pregnancy almost never survive. Their brains, lungs, and hearts are too underdeveloped to function outside the womb. Twenty-four weeks is generally considered the cutoff for fetal viability, meaning the point at which a baby can survive on its own, outside the womb. Every day the baby girl stayed inside, her chances of survival would rise.

As a trained physician, Kimberly understood the odds. And they weren't good. She tried to stay calm as she explained to Jamie what was going on. Kimberly was put on progesterone, a medication intended to prevent contractions. Per her doctor's orders, she continued to live life relatively normally for a few weeks: going to work, seeing patients (though she could no longer pick up the children for hugs as she loved to do).

At twenty-five weeks, in her sixth month of pregnancy, Kimberly checked herself into the hospital and was placed on bed rest. By then, the blood vessels in her baby's tiny lungs were still developing, not yet strong enough to pump on their own. Two weeks later, at twenty-seven weeks, seven weeks after her diagnosis, the baby decided it was time to enter the world. Penelope, a sweet little girl weighing only two pounds, five ounces, arrived crying and Kimberly cried, too. "She was this beautifully slick, bright pink baby and just had the most beautiful face—I'm totally biased, but she was so gorgeous."

The neonatologist quickly whisked Penelope from the obstetrician's arms and put her in an incubator, a plastic dome-like crib that regulates the temperature, humidity, light, and sound in what became the tiny, premature baby's first home. Kimberly, who'd been able to hold Penelope only briefly before she was taken away, wept uncontrollably, her emotions swinging from gratitude and joy that her daughter was alive and relatively healthy to anxiety and fear. To survive, her baby would spend the first critical part of her life in the neonatal intensive care unit, known in hospital shorthand as the NICU.

Penelope stayed in the NICU for 109 days—three and a half months—getting help breathing from a CPAP machine and supplemental oxygen and help eating from a feeding tube. Kimberly had recognized from the beginning that her presence and that of her husband would be crucial. But her dreams of being there for her daughter quickly crumbled. She did not have paid parental leave. When Penelope was born in 2017, Massachusetts did not offer it. The federal Family and Medical Leave Act provided for unpaid leave, up to twelve weeks, based on time in the job. Kimberly met the length requirement, but, for her, as for so many parents, surviving with no income for three months was impossible. She'd been in school and training for over a decade. She had no savings, only debt, mostly in the form of significant student loans. Those loans weighed heavily on her. Growing up, Kimberly had seen how financial instability could wreak havoc on a family's life. Even with her raft of impressive degrees and status as a doctor, the precarious financial situation of her childhood cast a long shadow. The anxiety it sowed had stuck with her. "Nothing felt secure," she says. "It scared me to pieces." Moreover, as the only Spanish-speaking provider at the clinic where she worked, she felt tremendous responsibility toward her patients, and didn't want to leave them.

And so, two weeks after giving birth, Kimberly was back on the job (like 25 percent of mothers in our country). She knew Penelope was likely to be in the hospital for a very long time; she planned to accrue sick days and any other paid days off so that once her baby came home, she would have guilt-free, paid time to spend with her.

Kimberly is far from an anomaly. The lack of paid family leave forces many parents into a desperate, unwinnable calculus. All babies need their parents, but perhaps especially those in the NICU. Studies have shown that premature babies whose parents spend more time with them in the hospital have improved brain and language development. They make up for time lost in the womb when they are held close to a parent's warm body.[8] And in 2021, preliminary results from an intriguing study showed that when parents had three months of paid leave as opposed to unpaid leave, their babies' brain function showed distinct EEG profiles, "possibly reflecting more mature patterns of brain activity."[9]

Kimberly was forced to make the same decision as so many others and continue working for financial reasons instead of taking care of her fragile newborn baby, even though as a pediatrician, she knew better than most *exactly* what Penelope needed. "She needed me to be skin to skin. She needed me to read to her. She needed me to sing to her. She needed me to be there in the hospital and advocate on her behalf," she says. Instead, at a time when her daughter needed her most, Kimberly had to go to work to take care of other people's children. She and Jamie spent as much time in the NICU as possible. Jamie's job offered two weeks of paid family leave, which he took. After those two weeks, he still spent every evening in the NICU, sitting there for an hour and a half with Penelope on his chest. Kimberly was there every night after work.

Being with her own daughter in the NICU was an eye-opener for Kimberly. Until Penelope was born, Kimberly had only ever experienced the unit from the doctor's side of the equation. "I never understood the trauma associated with being a NICU parent," she said.

But now, she feels something equivalent to PTSD about her months in neonatal intensive care. The experience also shocked Kimberly into rethinking some of her assumptions. As medical providers, we consider it standard procedure for parents to be involved in an infant's care. It's easy to pass judgment on parents who do not or cannot visit their children in the NICU. Kimberly had never considered that some families might be more privileged than others in this regard, until she experienced it herself. And she was angry, because this—parents not being with their babies in the NICU—felt like a problem that had a clear solution: paid leave.

Deeply frustrated at the disconnect between what she knew to be medically necessary and what she and every other parent were routinely offered, Kimberly felt a spark ignite in her. The fire of advocacy was lit. By then, when Penelope was ten months old, she and Jamie had moved to North Carolina. After meeting with advocates there, she collaborated with others to write an op-ed that combined her story with a call to action. Her status as a pediatrician and her pain as a parent infused her voice with authority and authenticity. Soon she was getting calls from journalists to talk about her experience and her policy propositions.

Next, she brought the issue to the American Academy of Pediatrics, a nationwide professional advocacy organization dedicated to improving pediatric healthcare standards. Every year, the AAP accepts suggestions from members on resolutions they would like to see adopted. Kimberly submitted an argument that the group should support legislation to promote paid family medical leave for all parents.[10] It is an issue clearly tied to children's healthy development—AAP's driving principle. "This is a no-brainer," she says. "Why would we not be [in favor]? This is what we do. We care for children and families." When a member submits a resolution, the academy's leaders vote on the top suggestions and then board members must make a strategic plan to address the top ten issues. Kimberly's was voted

number eight. As I write, she is helping the AAP craft a policy statement on the issue to help inform policy makers. And she's fighting
at the state level, too, in North Carolina.[11]

Fortunately, Penelope, who was three when I spoke with Kimberly, has done well. At nine months she no longer needed supplemental oxygen, and by the time she was two years old she was
finally able to stop using a feeding tube. Other than a bit of asthma,
she's healthy. Although she got off to a frightening start in life, Penelope appears to be thriving. Her growing brain—remember, it has
already more than doubled in size—is supporting everything she
can do, the skills she has mastered and those she is still working on.
"She's just . . . she's a very willful child," Kimberly says with a laugh.
A disposition not unlike her mother's.

The Essential Spark

I didn't get to meet Penelope, but I can easily imagine her at three,
telling her mother all about the friends she met at preschool, singing
her ABCs, or insisting on helping her mother stir the batter for
brownies. These are the kinds of behaviors that develop in those
magical, critical, first three years of life. Who was supporting the
brain development that makes those skills—and they *are* skills—
possible? Kimberly and Jamie . . . and the other loving caregivers
who helped along the way.

The circuitry in children's brains develops based on the input
they receive. Who delivers that input? Parents—and, in some instances, older siblings, nannies, grandparents, family friends, childcare providers. They are the brain architects. As I discussed earlier,
the essential spark fueling those trillions of neural connections that
form in children's brains is the serve-and-return interaction, the
back-and-forth conversation that occurs between caregiver and

child. Neuroscience tells us that these interactions, which look like very simple banter between adult and child, provide critical neural nutrition to the developing brain.[12] Begun on the very first day of life when the parents first coo and cuddle with their baby, and continuing through preschool, this vital verbal serve-and-return nourishes and stimulates the brain during this uniquely fertile period and gives it the chance to reach its ultimate potential.

These foundational ideas inspired me when I started the organization that became the TMW Center for Early Learning + Public Health. From the start, my thinking and that of my team was spurred by research that showed significant variation in the quantity and quality of words that individual children heard and subsequent meaningful differences in language and academic outcomes. We wanted to apply that research in order to level the playing field. And we knew, because it has been clearly and scientifically demonstrated, that babies aren't born smart, they're *made* smart. Intelligence can be and is shaped by experience and environment. It is malleable. Since it is parents who do that shaping through serve-and-return, everything we planned to do at the TMW Center would be parent-centered, parent-tested, and parent-directed. The first important question we wanted to address was how to help parents achieve an optimally rich language environment for their children. How could we make serve-and-return interactions a natural, no-fuss part of a family's everyday routine?

Our answer—the 3Ts: Tune In, Talk More, and Take Turns. This became our core strategy. Tuning in encourages parents to make a conscious effort to notice what their baby or child is focused on, and then to talk with the child about whatever that is. Talking more adds words to the piggy bank in a child's brain—a piggy bank that earns compound interest. The more words put into the bank, the more brain connections a child builds and the bigger the child's vocabulary becomes. Taking turns is all about conversational exchange and active engagement. It becomes a social dance of sorts. Early on, before children can speak, when parents give their children a turn at weighing in, every coo, clap, and cheeky grin is a turn. It might not look like conversation, but science shows that each interaction is a key building block to growing the child's brain bank.[13]

Remember the story of Mariah playing with her son, Liam? That was the 3Ts in action, and it was something she, like many parents, knew instinctively how to do but that was bolstered by working with us. As she began to read the book about Chomp the lion and Liam took the book from her, wanting to be in charge, she let him take the lead because she was *tuned in* to what Liam wanted to do. When she picked up another book, *Ten Little Fingers and Ten Little Toes*, that gave her a chance to *talk more*. "There was one little baby that was born in the hills, and another who suffered from sneezes and chills," she read. "And both of these babies, as everyone knows, had ten little fingers and ten little toes." When she then grabbed Liam's fingers and toes, this helped him to make the connection between the words she spoke and his own body. Mariah also made sure to *take turns* in the "conversation" they were having. At the end, once he had flipped through all the pages of the book he was holding, she said, "Good job!" When he echoed her words—"Job!"—that was Liam taking his turn.

The 3Ts are catchy and easy to understand. They also capture exactly what parents and caregivers can do to help young brains make new connections. They translate the complex science of language exposure and brain development into an accessible, easily adoptable program that enhances everyday parent-child interactions. And they provide everyone—parents and caregivers, grandparents and babysitters, no matter their own academic attainment, no matter their wealth, no matter their jobs—with the essential strategies for optimally building a child's brain. The best part?

Simply talking, tuning in, and taking turns with your child helps to ensure your child's ability to achieve the promise of their inborn potential.

In the decade I've been doing this work, however, the science has not stood still. It never does. Now we know much more about how and why, when all goes right, a rich language environment works its magic. But more important, we understand with enhanced clarity what is not happening in the brains of children whose worlds provide less language, less conversation, and less engagement. The latest science only strengthens the thinking behind why the 3Ts work and reinforces the crucial role of parents.

Until recently, neuroscientists could study only one brain at a time.[14] They could look at the activity in a mother's brain while she gazes at photographs of her baby, for instance, and see that the emotional parts of the brain light up (of course, they do!). Or they could put an EEG cap on a baby and measure how quickly her brain waves respond to hearing a story. But they could not do both at the same time—they could not capture and compare activity in the mother's brain while she was reading *The Runaway Bunny* with activity in the baby's brain as she was listening. Now they can, thanks to a cutting-edge technology called hyperscanning, which measures activity in two (or more) brains while they interact. When a group of people attends a Rolling Stones concert, for instance, swaying and singing along to "(I Can't Get No) Satisfaction," it turns out their brain waves come into alignment.[15] This phenomenon is known as neural synchrony, and it means that as individuals pay attention to the same thing, the patterns of electrical activity in the brain become more and more similar. And when neuroscientists had one person tell a story and the other listen, the level of the listener's comprehension was much higher when the pair's brains showed neural synchrony.[16]

Those findings were especially intriguing to scientists who study how babies acquire language. Could they see language learning in action in the interplay between adult and child? What would neural synchrony between a mother and baby look like? What would it mean? Now, let me say, it is very hard to study the brains of babies. They squirm, they pull off the monitoring caps, they fuss when they're tired. But the neuroscientists at the Princeton Baby Lab are very good at enticing babies to cooperate. In a recent experiment that highlighted the importance of *tuning in*, they measured the degree of neural synchrony between babies and a playful adult.[17] Eighteen babies between nine and fifteen months of age came into the lab. In each session, both the baby and the female experimenter wore caps studded with monitors that measure the level of oxygenation in the blood, an indirect way of assessing brain activity because metabolic activity demands more oxygen. (The technology is called functional near-infrared spectroscopy, or fNIRS.)

In one part of the experiment, the "together" condition, each baby sat on a parent's lap and played with the experimenter as she brought out toys, sang nursery rhymes, and read *Goodnight Moon*. The experimenter was all smiles and fun and spoke in the singsong voice of child-directed speech or "baby talk," which we know appeals to babies. At other times, in the "apart" condition, the baby sat in the same spot, but the researcher told a story in adult tones to a fellow experimenter in the room. The baby could hear what she said, but the conversation wasn't directed at the baby. In essence, this was overheard speech. The results were simultaneously predictable and profound. The brain waves of the baby and the experimenter were significantly more in sync when the two were directly interacting than when the adult's conversation was directed elsewhere. Furthermore, the parts of the brain involved in mutual understanding showed greater activity when the adult and baby were gazing at each other and when the infant was smiling.

Tuning in puts caregivers and babies on the same wavelength— literally. "While communicating, the adult and child seem to form a feedback loop," said Elise Piazza, the Princeton Baby Lab neuro-scientist who led the study. "That is, the adult's brain seemed to predict when the infants would smile, the infants' brains anticipated when the adult would use more 'baby talk,' and both brains tracked joint eye contact and joint attention to toys. So, when a baby and adult play together, their brains influence each other in dynamic ways."[18] Neural synchrony may one day be something we can use as an indicator of successful coordination in mother-child conversation. It explains why baby talk is good for babies! Other studies have shown that neural synchrony boosts social learning, problem-solving skills, and the ability to learn new vocabulary. Tuning in by an adult is a critical first step that primes a baby's brain for the learning that lies ahead.

There is also exciting new evidence for why *taking turns* is such an essential part of early language and brain development. That evidence comes from the work of Rachel Romeo. Around the time that we at TMW were developing the 3Ts, Rachel was in graduate school, working toward a PhD as a speech-language pathologist and a neuroscientist. When I first met Rachel at a conference, her intelligence and passion were clear, and I guessed right away that she was going to do great things.

Like me, Rachel wanted to understand—and, importantly, do something about—the disparate outcomes that seem to go along with socioeconomic status. Also like me, she was inspired by the early research on the significant effects of early language exposure. But Rachel wanted to go deeper than the existing research and understand exactly what that language exposure was doing in the brain that had the power to change a child's trajectory. "I'm certain it gets under the skin," she says.

In a first-of-its-kind study published in 2018, Rachel and her colleagues at Harvard and MIT (she's now at the University of Maryland) put thirty-six four- to six-year-old children in a brain scanner to look at brain structure and brain function while they listened to stories.[19] Then she sent the children home with audio recorders and asked their parents to record everything the children heard for two days, from the moment they woke up in the morning until they went to bed at night. When all the recordings came back, Rachel crunched the numbers and compared the data on language to the brain scans. The benefits to the brain in the children who experienced more conversational turn-taking were obvious, and what's more, they registered as important over and above the sheer volume of words heard. That meant the quality of language—that it was directed at a child and part of a lively serve-and-return interaction—mattered even more than the quantity of words heard.

The differences in those children who did the most conversational turn-taking were evident in two important ways. When they listened to the stories while in the scanner, there was more activation in key language areas of the brain—they "lit up" more. And there were stronger connections *between* language areas that govern speech perception and speech production, what you hear and understand and what you say. "These kids who were experiencing more conversational turns had much more robust connectivity in those areas," Rachel says. "Their brains seem to be maturing a little faster." In other words, kids who got more opportunities to talk with their parents had a real advantage in the efficiency of their brains. That is what increasing connectivity means, that you are physically building stronger, more streamlined brain circuitry. When she went a step further and did a statistical analysis of language development, Rachel found that the same kids were progressing faster in both vocabulary and grammar.

Here's the really striking thing: The children in Rachel's study represented all parts of the socioeconomic spectrum. And she found that no matter what kinds of families they were from, the amount of conversation turn-taking they experienced mattered more than

income level. It was the top predictor of the changes in brain structure and function and of stronger language development. "Even
when you're surrounded by adversity on the exterior, if you have a
nurturing home environment you can buffer a lot of those adverse
effects on the brain," Rachel says. In theory, this is something every
parent can control. In practice, external circumstances often get in
the way, particularly if there is simply no time available for this kind
of conversational exchange.

Rachel's study has important implications. It drives home the
truth of what we know about the power and possibility of adult
and child engagement in the early years of life. But what is it about
taking turns that supports such critical development in the brain?
She and her colleagues are still trying to figure that out. She suspects the personal connection that is forged between adult and
child has a lot to do with it. "Conversation means that it's a two-way
street," Rachel says. "You're engaging with each other. You're experiencing language but you're also producing language. I like to think
of it as creating a feedback loop." (It's no accident that Rachel uses
the same phrase—"feedback loop"—that Elise Piazza used. It gets at
the heart of the interactivity that parents and children must establish.) Engaging in conversation has another advantage, too. It allows
parents to assess their child's language level and to pitch their own
conversation there, where they're sure the child can receive it. "We
call it the zone of proximal development, which is really that sweet
spot where you're talking right at their current development level,"
Rachel says.

Finally, Rachel tested these findings in a nine-week intervention
designed to increase conversational turn-taking. The children who
participated, again four- to six-year-olds, showed positive changes in
verbal, nonverbal, and executive function measures (such as following directions and showing cognitive flexibility), as well as structural brain changes that support language and social processing. As

Rachel's colleague at MIT, John Gabrieli, put it, "It's almost magical how parental conversation appears to influence the biological growth of the brain."[20] We could see this, too. Early on at TMW, we developed curriculums designed to bring the 3Ts to families in their homes, educating parents on the importance of tuning in, talking more, and taking turns. Our goal? To help parents embed the 3Ts in their everyday interactions with their children. When we tested the effectiveness of the program against a comparison intervention (it involved instruction about good nutrition), we saw that over the course of a few weeks, the children of parents who learned the 3Ts showed improvements in language development.[21]

But would the 3Ts make a difference over the long haul? How much could the 3Ts change the trajectory that children were on? To find out, we embarked on an ambitious longitudinal study.[22] We have been following children and their parents—both English and Spanish speakers—who participated in our Home Visiting Program from the time the children were around fourteen months old all the way into kindergarten. (We started following the Spanish speakers when they were toddlers.) We are tracking progress in the children and changes in the parenting styles of their mothers and fathers. Family after family, home after home, we have gotten to know parents and their children from infancy through kindergarten. Over and over again, I have been awed by the raw determination of parents to give their children a better life. No matter how hard my team and I worked, the parents worked harder.

Home visiting was as educational for us as I think it was for these families. It gave us a window into their lives. What we saw most clearly was their commitment to using the 3Ts all day, every day, in spite of—or perhaps because of—what must have felt like insurmountable odds. I began to see that the 3Ts had to work as part of something larger, as part of day-to-day lives in which parents have not just the knowledge but also the space and time—made possible by reasonable employment policies and access to high-quality childcare—to make the most of the gift that evolution has given them.

A Life Interrupted, a Child Cheated

Of all the families we have met at TMW, perhaps none faced worse odds than Michael and Keyonna. They had been together for two years when she became pregnant with their child. They were both thrilled. Michael dreamed of creating a life and family with the confident, vivacious woman he had had a crush on long before he got up the nerve to tell her how he felt. And Keyonna knew that Michael, who was quiet, steady, and soft-spoken, her gentle giant, would be a terrific father. With her two sons from previous relationships, Cash and DiaMonte, he was a loving and involved father figure. He helped teach Cash his colors by running through the crayons in a Crayola box and he taught him his ABCs. Michael also regularly walked the boy to preschool, just down the street from their apartment.

In the middle of May, two months into the pregnancy, Michael took Cash to school as usual. He signed Cash in and retrieved some homework from the boy's cubby. *I'll help him with this tonight*, he thought. He had just left the school, homework in hand, when the young family's life was abruptly turned upside down. A large SUV suddenly pulled up in front of Michael, blocking his path. Three Chicago cops were inside. "Come here," they said. Michael was frightened but didn't want to make things worse. He was respectful, but then two more SUVs pulled up. The cops piled out and at least one pulled a gun. They put Michael in a car and took him to the station. He remembers there was an Aerosmith song playing on the radio and that one of the cops said, "Listen and enjoy it. It's the last music you'll hear."

Meanwhile, when Michael didn't come home, a concerned Keyonna, who had been making them breakfast, started calling relatives and friends, but no one had seen him. She went to the school and saw that Michael had signed Cash in. Finally, she started calling police stations. On the third try, she was told that Michael was being held for murder.

Six years earlier, a convenience store worker had been shot during a robbery in Michael's old neighborhood. The cops pegged Michael, who is Black, as their main suspect based on two questionable witness identifications. (One of the witnesses described the perpetrator as about five feet nine even though Michael was nearly six feet four. The other witness later recanted his identification of Michael and couldn't pick him out in court.) The police also said they had DNA from the scene of the crime that would match Michael's. He was terrified. But he knew he hadn't done it and he believed this case of mistaken identity would soon become obvious.

He was right about the obvious part. When Michael got his day in court, the jury quickly found him not guilty. But that day in court came not five hours or five days or even five months later. It came five years later. Despite Michael's Sixth Amendment right to a speedy trial, it took four years to set a trial date, and then Michael's public defender had to delay, first because of childcare issues and then the death of her daughter. Then the lawyer herself passed away and another attorney had to take over his defense.

Michael had been denied bail and spent those five years in prison. (Rates of pretrial detention have risen dramatically over the last twenty years even though the U.S. system presumes innocence until someone is proven guilty.)[23] Michael and Keyonna's son, Mikeyon, was born six months into the ordeal. Throughout the next four and a half years, Keyonna brought Mikeyon to the prison for visits, but Michael never once got to hold his son. He barely got to talk to him.

I got to know Keyonna after she signed up for the TMW home

visiting early language program and became part of our longitudinal study. Despite the enormous stress of parenting three children and worrying about Michael, she was determined to do everything she could to give Mikeyon the best start in life. But what she couldn't do was change the fact that for those sixty months, before Michael finally got a trial and was quickly exonerated, Mikeyon was deprived of half of his parental interactions. Interactions that we take for granted, like the oxygen we breathe, until they are suddenly taken away.

What happened to Michael was unjust and unfair in every way. But it was also unjust and unfair to Mikeyon. He paid a price we rarely calculate when we consider the repercussions of the criminal justice system.

Perhaps Michael and Keyonna's story feels extreme, a dramatic and horrifying turn of events, and not one that is likely to happen to many parents. But our nation has a history of separating children from their parents, a practice that demonstrates callous disregard for the intense bonds between parent and child, for the sustaining power of loving caregiving, and for parents' role as brain architects. I see their story as a stark reminder of the barriers—of all sizes—that are thrown up in front of loving parents.

A Thought Experiment

As I thought about the many families I worked with, I found myself imagining a thought experiment. A "what if" for modern parents. It's based on a similar exercise suggested by the political philosopher John Rawls. His instruction to students was this: Design a future society in which they would want to live. The catch is that you have no idea where you will fit in that society. You might be rich or poor, blessed with great intelligence or not, and so on. You must design from behind what Rawls called "a veil of ignorance."[24] This changes things. I once read a moving article by the writer Michelle Alexander which asked us to imagine, as Rawls did, that we were all reincarnated. Alexander put the implications well: "If we're born again at random, we can't soothe ourselves with fantasies that we'll come back as one of the precious few on the planet who live comfortably . . . What kind of political, social, and economic system would I want—and what would I fight for—if I knew I was coming back somewhere in the world but didn't know where and didn't know who I'd be?"[25] The exercise makes it awfully hard not to acknowledge both the lottery of life and the extent to which the destiny of any one human is linked to that of all the others.

In my version of the thought experiment, the question is not what if you were reincarnated as someone very different in the future but, rather, what if you gave birth to a child but knew that someone else would raise that child? I'm not trying to explore the misogynistic horrors of *The Handmaid's Tale* here. Or telling you to believe in reincarnation. Rather, I want to force us all to think deeply about what it takes to raise a child well, because I know most parents will do anything to ensure that our kids are okay, even in a thought experiment. Put yourself on that riverbank with a line of parents on either side of you and recognize that any one of them might be the person who will have to steer your child across the raging torrent of water. In this imagined scenario, you have no control over who cares for your child, who navigates. On the other hand, you *can* plan for the supports and safety nets that the future society provides—for the boats, in other words. What would you want every parent to have to make sure that, no matter who ends up raising your child, the child will be adequately cared for and educated? How would it change your priorities if whatever comforts you currently enjoy and can give to your child were no longer certain at all? How would you make sure that every child had what they needed to get off to a strong start with optimal brain development, and every parent had what they needed to help with that task? And while I'm at it, what would it take to raise that child for two decades and to launch them into an adult life as a fulfilled and productive citizen?

It's a sobering thought. You begin to see how everything from paid parental leave to a more just (and speedy) criminal justice system would become important. What would Charlotte and Hazim need? All that speech therapy for Charlotte, for a start. Access to good schools in his own neighborhood for Hazim would be essential. What would Mariah want? A high-quality childcare system for her children and one that she herself could work in for a fair wage and with chances for professional development and advancement. What about my three children when they lost their father? There are some tragedies we can't fix. But I could not have raised them without considerable support from the family and friends who stepped in after Don died, as well as the paid caregivers I could afford to hire. (Our nanny, Lola, was like a second mother.)

You don't have to believe in reincarnation to imagine a world that you'd want your child to be born into no matter where and by whom they would be raised. You do have to fight for that world. You do have to recognize the critical importance of parents and loving caregivers doing the essential work of building young brains. And I know that the world I imagine would be far easier to accomplish if we could all agree on healthy brain development as our priority, and on parents as the way to achieve it.

• PART TWO •

THE
DISCONNECT

FIVE

IT ALL STARTS WITH BELIEFS

"It isn't enough to talk about peace. One must believe in it.

"And it isn't enough to believe in it. One must work at it."

—ELEANOR ROOSEVELT[1]

e are not born doctors, teachers, drivers, or engineers. To pursue any of those professions, we go to school to study and learn. We are not born parents either. But there is no school to teach us parenting skills. We are left alone to figure out what to do in this hugely important endeavor.

In 1999, when I had my first child, I probably read a dozen parenting books, feverishly scribbling in the margins as if I were prepping for board examinations. My late husband, Don, and I attended parenting classes, bought a top-of-the-line car seat, and made sure our baby had the best and safest of everything. We were career surgeons in constant contact with children—he a pediatric surgeon working with kids every day, and I a pediatric cochlear implant surgeon working just millimeters from children's brains. We were trained to plan for every contingency. Everything, it turns out, but the one thing we most needed to know about—as parents, that is. We knew nothing about healthy brain development.

Sure, we knew that a baby's environment made a difference. That we should nurture and love our baby. But somehow, I didn't connect the dots between everything I was reading and the brain that my baby girl, my sweet Genevieve, would develop. I think that's because it was never laid out in plain English that the way I interacted with my baby, the way I spoke to her, would form the connections she needed in her brain. Not in medical school. Not by my obstetrician. Not in our pediatrician's office. Until quite recently, parents were never told that they hold within themselves the ability to build a child's brain.

By the 1990s, when neuroscientific knowledge exploded, parenting books did talk about building children's intellectual and language skills, but they rarely linked those ideas directly to foundational brain development. As journalists say, they buried the lede by not pointing out just how much the conversation between parents and children contributed to wiring up kids' neural circuits. We tended to conclude that educational products—like Baby Mozart videos—were the answer to developing our children's brains. The truth is that, all along, the answer lay inside each of us. It lay in my ability just to tune in to Genevieve, to talk to my baby, and to show her from day one what conversation and interaction are all about.

In the twenty years since I had Genevieve, everything has changed—and nothing has changed. As a society, we know so much more about brain development. The science I've already described makes that clear. Yet we act as if parents are born with this information. We still don't connect the dots for them, and we fail to give them explicit information about how the brain develops or what tools they can use to encourage and enhance it. We leave it to them to figure out what to do. The result is a country full of anxious parents and a frenetic parenting industry rife with misinformation and conflicting advice. Yet what parents know and what they believe is critical. They need to believe they can build their children's brains. Because they can!

Now, to be clear, the impulse to nurture and protect our children, which is at the heart of what is required, is embedded deep within our DNA and our brain circuitry. Becoming a parent looks a lot like falling in love—and it looks that way in our very neurons, because the neural changes that occur when we are newly infatuated with a romantic partner are similar to those that occur in the first months of being a parent.[2] Falling in love with our children is nature's way of ensuring the care of those helpless newborns who arrive, as I explained earlier, "underdone." Falling in love encourages us to make the most of this marvelous evolutionary gift of extended brain development.

"Wait until you become a parent," my mother used to say when I would complain that she was overprotective, as if everything would be different once I had kids. And it is. Now I say it to my own children because there is a fundamental truth to those words (as well as a universality to children thinking their parents are wrong). Women's brains change structure when they give birth. The neuron-rich gray matter becomes more concentrated. Regions that contribute to empathy, anxiety, and social interaction become more active. These changes, spurred by a flood of hormones during pregnancy and right after giving birth, help bond a mother to her baby. Multiple studies have shown that just staring at her baby is enough to set the reward centers in a mother's brain firing. And her brain responds differently to photos of her own smiling baby than to other babies.[3] Fathers undergo hormonal changes, too, a biological reflection of a shift in goals—from reproducing to raising children.[4]

If our primal desire to love, protect, and get our kids to the other side, to launch them successfully into adulthood, is so deeply embedded, why do we need further instruction? Because what it means to parent well has changed over the millennia. For thousands of years, the definition of success for a parent—what it meant to cross the torrent and reach the other side of the river—was simply to keep their child alive. It was quite literally an eat-or-be-eaten world. Evolutionary biologists believe that the need to protect their young led to an ability to "tend and befriend" among females in particular, because nurturing and getting support from others was the way to save your kids from the actual lions thousands of years ago and from the figurative lions today.[5] In time, the risks came less from dangerous beasts and more from dangerous diseases. The Sewalls, the Puritan family we met earlier, lost seven of their fourteen children—a devastating level of loss that was horrifyingly common in that era. Well into the early twentieth century, infectious disease was the world's predominant killer, accounting for about one-third of all deaths, a bit over 30 percent of which occurred among children under five.[6] Accordingly, advice to parents focused on hygiene, nutrition, and physical health. Into the 1930s, it was rare for pediatricians to present babies as "thinking, learning, curious beings," wrote historian Julia Wrigley. A handful of experts went so far as to recommend to parents that they avoid stimulation entirely because the baby's growing brain would be harmed by it.[7] John B. Watson, the president of the American Psychological Association, included a chapter called "The Dangers of Too Much Mother Love" in a 1928 bestselling book on childcare.[8]

Once vaccines, antibiotics, and improvements in hygiene and sanitation came along, deaths from infectious disease—and particularly deaths among children—dropped dramatically. By 1997, less than 5 percent of deaths were due to the major infectious diseases (pneumonia, flu, and HIV, by that point), and young children accounted for less than 2 percent of all deaths, including those from infectious diseases.[9] As a result, parents could put their attention on the long-term outcome and establish different priorities: They could take good health for granted and think about how to foster their young children's intellectual growth.

By the second half of the twentieth century, the importance of babies' cognitive development had become common knowledge, even if a lot of us (guilty!) didn't fully grasp the big picture. Interest in cognitive development first took off in the 1960s. It was increasingly understood that babies need nurturing, appropriate stimulation, and that babies learn from every touch, every word, every sound, and every smell. In addition, parents themselves got markedly better educated as the twentieth century progressed, and it became more common to finish high school. The more educated the parent, the more likely they are to engage with their children's cognitive, social, and emotional development.

Something else has shifted as well: our ideas of what good parenting looks like. For decades, middle-class parents have tended to pursue an approach the sociologist Annette Lareau described as "concerted cultivation." As the name suggests, that's a style of parenting that is child-centered, active, and involved. It's also known as "intensive parenting" (or, when it is being mocked or criticized, as "helicopter parenting"). By contrast, in the past, working-class and less-educated parents appeared to favor a more relaxed style of oversight dubbed "natural growth."[10] But more recent research reveals that the story of class differences and parenting preferences is more nuanced than that, in two important ways. First, some of those previously documented class differences in parenting probably reflected differences in resources more than in ideas of what constitutes good parenting. Intensive parenting takes time and money. (Culture, race, and ethnicity probably also played a role.) Second, where once there were substantial class differences in parenting views, they seem to have largely melted away. Parents across the socioeconomic spectrum now espouse remarkably similar ideas about what good parenting looks like, says Patrick Ishizuka, a sociologist at Washington University. They all lean toward intensive parenting.

I called Patrick to learn more about a fascinating study he conducted.[11] (I was amazed that as the father of a two-year-old and five-month-old twins, he found time to talk to me!) He asked more than 3,600 parents across socioeconomic classes to read descriptions of parenting in common situations involving eight- to ten-year-olds. How should parents deal with a child who claims to be bored? How should they respond to a kid who's acting out? And how involved should parents get with a teacher who gives bad grades? In each vignette, the parenting response could be classified as either "concerted cultivation" or "natural growth."

In all the parenting situations Patrick created, no matter the gender of the parent or child, about 75 percent of parents—both those with and without a college degree—thought the concerted cultivation responses were excellent or very good, whereas just over a third rated natural growth parenting that way. It wasn't that parents said the natural growth parenting was bad or equaled poor parenting. Many thought it was "good" or "OK." But, Patrick told me, "both more- and less-educated parents viewed the intensive parenting as ideal."

In reality, few people's circumstances allow them to fully achieve this "ideal." For some, it is virtually impossible even to make a start. Yet that hasn't stopped intensive parenting from becoming the norm that most parents aspire to. They view it as the optimal way—maybe the only way—to ensure that their children will succeed in school and therefore in life. Intensive parenting is a logical response to a world plagued by income inequality and societal inequities. Though we like to think the United States is a country where anyone can succeed, a child born into poverty is likely to stay there. Now that it is harder to earn a comfortable living with a blue-collar job, education is considered the primary way to escape. And the return on investment from education is almost certainly driving parents' vision of how best to raise their children because the stakes are so much higher than they used to be.

Of course, middle-class and affluent parents are driven by some of the same reasoning, which partially accounts for the egregiousness of rich parents who are willing to break the law to get their children into select universities (as in the Operation Varsity Blues case).[12] Such actions only exacerbate the tremendous inequalities that already exist. No matter what we want for our kids, society has made it increasingly hard to get, and that makes intensive parenting seem like the only answer.

What Parents Know, What Parents Believe

I didn't really question my parenting in my children's early days—though I do often joke that I wish I had known then what I know now. By the time I began my journey outside the operating room, however, I had begun to wonder how the mighty brain architects—parents—come to know that they are capable of building a brain when it is quite clear that children don't come with instructions. In time, through our work at TMW, I came to understand that before we do anything else, parents must come to believe the essential ideas that inspire our work: that strong brains are made, not born; that they—parents and caregivers—are the architects who build those brains. They are the people who will guide us to the North Star.

Parents and the other loving adults in children's lives must not only know this, they must believe it. What's the difference? Knowledge is awareness, understanding, and a recognition of facts. Beliefs represent conviction that what you know is true; belief places trust in knowledge and accepts it on a level deeper than mere knowing. I *know* that two plus two equals four. I know and *believe* that the way a parent talks about numbers and patterns shapes a kid's spatial reasoning abilities. Knowledge and beliefs are a brain architect's instruction guide. They are the Cartesian coordinates leading to the North Star of healthy brain development.

Together, knowledge and beliefs shape a parent's behavior. If parents believe that intelligence is fixed at birth, they are likely to make choices about how to "invest" their time that are different from the ones they would make if they believed that what they do will impact their child's potential. For example, if a father doesn't realize that getting down on the floor, face-to-face with his baby, mimicking and introducing sounds, and eliciting responses—in other words, engaging in serve-and-return—is going to greatly impact his baby's intellectual growth, he might not bother. If a mom thinks spending that extra ten minutes playing blocks with her six-month-old isn't any better for the baby than a quick kiss on her way to take the laundry out of the dryer, that's an investment she is less likely to make. But when that mom *knows and believes* that those extra ten minutes make a difference, I'd bet just about anything that, even if she can't skip the laundry entirely, she will find ways to talk and sing to her little one as she folds the shirts. And Dad will be belly-down on the floor making funny noises for his kid. (The same is true when parents know that very young children learn from people and not from screens.)

What makes me so sure? TMW studies have found a direct relationship between a parent's knowledge and beliefs about brain development and how much they interact with their child from day one.[13] We see every day how information and understanding can transform parental interactions in a way that makes a huge difference in brain development.

Early on, I didn't fully appreciate just how much beliefs mattered. I suppose I thought they came along, more or less naturally, with the information we would deliver in our programs. But the more we talked with parents and followed their children's progress, the more I realized I needed to linger in this middle ground of knowledge and beliefs. You can't just share information with parents about how to talk to their children. You have to put time and energy into helping parents see that doing so has the power to build a brain.

I also discovered that there was no reliable way to assess what parents know and believe. Asking questions is not quite as simple as it sounds. You need to be sure that the question is phrased in a way that elicits the kind of information you seek. That's why researchers use validated and standardized sets of questions—measurement tools that are carefully designed and tested. But there were very few such tools that addressed the questions I was interested in: Did parents understand how children's brains develop? That it isn't a question of nature *or* nurture, but that nature must be nurtured? Did they understand all the ways early input grows the brain's capacity for everything from self-regulation and empathy to mathematical reasoning and reading? What did parents know about how the first years predict a life's trajectory?

To find out, my talented team and I created the Scale of Parent/Provider Expectations and Knowledge. We call it the SPEAK. It includes questions that deal with all the different aspects of a child's brain that a parent can have an impact on: cognitive capacity, literacy, math and spatial learning, social and emotional skills, executive function, and more. Building the SPEAK has been a huge undertaking, and we're not done yet. As I write, we're developing a more sophisticated computer-adaptive version of the survey that reduces the number of questions participants must answer while increasing its accuracy and precision.

The scale asks questions like these:

Is educational TV good for learning language?
Is there language development in the first six months of life?
Will being bilingual confuse a child?
Is a baby's brain born or built?

We found that how parents answer that last question is especially influential.[14] Some parents believe a child's intelligence is set at birth and remains virtually unchangeable by a parent or a caregiver.

Others believe in a child's intellectual malleability; that is, that a child's brain is built and that *they*, the parents or other loving adults, are the essential factors in ensuring its optimal development. The difference is everything.

The SPEAK let us pull back the curtains on what people really know and believe. It allowed us to see parenting through their eyes. For example, we went into a major birthing hospital and met with hundreds of moms from all different backgrounds—rich and poor, Black and white, English-speaking and Spanish-speaking—the day after they had given birth. Granted they were a bit bleary-eyed, but we found that many parents consistently underestimate how early children can be affected by critical experiences. In effect, there is a "missing" first year. Many are unsure of the value of child-directed speech from birth, of sharing books with infants, and of the way their own interactions with their babies will build language development. More than half thought that infants can learn language from television and screens (they can't). What was incredible to me was that we didn't find that there was much learning by doing: There was surprisingly little difference between the knowledge of first-time parents and of veterans.

Other researchers are asking similar questions and getting similar answers. In a large 2015 survey of more than 2,000 American parents, half mistakenly believed that quality of parent care has a long-term impact only after babies are six months or older. A majority didn't believe that babies have the capacity to feel sad or fearful until they are six months old, when in fact that happens at three months. Nearly half said that reading to children starts to benefit long-term language development about a year and a half later (at two years plus) than it actually does (at about six months). And the majority of parents didn't recognize that the benefits of talking to babies begin at birth.[15]

Our work also painted a consistent and compelling picture:

Although nearly all parents value intensive parenting, money and education do make a difference. Parents with more of either knew more about how their own investments of time and energy affected their children's development. And every bit of extra schooling increased the effect so that the results follow a clear gradient. Parents who didn't finish high school knew the least, followed by those with only a high school degree, then those with some college, and finally those with a college degree or more, who were best informed.[16] It's important to emphasize that those are averages, and there was huge individual variability at every income and education level, meaning that some low-income parents who hadn't finished high school knew more, and some more affluent parents who had college degrees knew less, than their peers.

To dig deeper into these differences, we met with families from lower-income backgrounds and, again, saw dramatic variability. But we didn't just confirm the wide range in knowledge levels, we also began to demonstrate the way parental knowledge maps onto parents' interactions with their children. We followed nearly two hundred families during the first year of their babies' lives. We found that, as expected, what parents know as early as the first week of their child's life predicts what those parents do with their children as the babies grow. And we also found that those parents who knew more interacted with their babies earlier and more often. Specifically, parents with lower incomes who knew more about cognitive and language development in the first week were more likely than other such parents to respond sensitively to their babies, to engage with them and, therefore, to foster both their cognitive and their social and emotional growth. This suggested to us that if we could increase parents' knowledge—and beliefs—about their children's language and brain development, we could have a positive effect on their parenting skills no matter how much education a parent had.

To find out if this was true, we launched two related studies designed to assess whether increasing parents' knowledge would change what they did.[17] In one, we talked with more than four hundred parents while they sat in the clinic for four well-baby pediatric visits (at one, two, four, and six months). Each time, we showed them a ten-minute video explaining what it means to tune in, talk more, and take turns with a baby. The videos were full of examples of mothers and fathers using the 3Ts while snuggling, diapering, reading books, and so on.

In one, a dad snuggling with his newborn says, "You know basketball's in your blood. I'm going to show you the signature jump shot."

Or a mom talks her baby through their day while nursing. "You are hungry, aren't you? After this, we'll have a bath, get some new clothes to wear . . ."

In a second, more intensive experiment, ninety-one parents (all Spanish-speaking) agreed to have us come to their homes for twelve hour-long monthly visits during their children's third year of life. Half served as a comparison group, and they received information on healthy eating and nutrition instead of information about brain development. For the other half, each visit included an educator who shared the science on a specific developmental topic like giving encouragement or incorporating talk about math into everyday routines, and then gave advice on how to implement such strategies. The educator led parents and kids in activities that put the science they'd just talked about into practice. Parents got down on the floor with their kids and tuned in, talked more, and took turns. Our home visitors also provided feedback and helped parents set goals.

Parents soaked up the science behind their children's brain development. Both of our studies showed that our interventions strongly affected parents' beliefs and increased their interactions with their kids, resulting in more of the precious serve-and-return exchanges that are so critical for babies' brain development. Not surprisingly, the intensive home visiting program was more effective than the newborn videos because it was one-on-one, and it not only provided knowledge and strategies for brain development but also showed how to put those strategies into action. In the home visiting program, there were improvements in children's vocabulary, math skills, and socioemotional skills. The strongest link was for linguistic skills, which isn't surprising either, given how much we emphasize the role of parent talk. Put simply: Knowing more affects how parents behave with their children, and enriching parent behavior changes children's outcomes.

Believing in Math Talk

How often have you heard someone say, "I'm not a math person"? Far too often, I would guess. How we talk about math and our beliefs about the subject have been getting a lot of attention in recent years. Girls and women are particularly susceptible to the belief that they aren't good at math, and beliefs about math ability are reflected in performance.[18] Belief can become a self-fulfilling prophecy.

It was through thinking about math talk and what we could do to improve the SPEAK that I got to know Talia Berkowitz. Talia had been a graduate student in psychology and then worked as a postdoctoral fellow in the lab of my friend and colleague Susan Levine, the same researcher who followed the progress of Charlotte, the girl born with half a brain.

I needed Talia's expertise to help us build out the math domain in the SPEAK. I was sure she could help us because she had cut her teeth on the intricacies of math and spatial development while working with Susan on an assessment of a digital program called Bedtime Math. Bedtime Math is an app designed specifically for parents to use with their children as part of their nightly bedtime story routine. Parents don't always realize that talking about patterns, categories, and comparisons all relate to math and spatial skills and help build their children's brains. The math-rich stories in Bedtime Math make it easy. For instance, a short story about an octopus taking a picture with a waterproof camera is followed by questions about the octopus's eight arms, beginning with "Who has more arms, you or the octopus?" for the youngest kids and getting tougher from there. But to be sure Bedtime Math was working as intended, its founders asked Susan and Talia to investigate. Did the app have an impact on children's learning?

They designed a rigorous study in which parents of nearly six hundred first graders used the app throughout a school year. Their findings were published in *Science*, one of the most prestigious academic journals.[19] Sure enough, Susan, Talia, and the team found a significant improvement in the math abilities of children who used Bedtime Math regularly with their parents, compared to kids who used a reading app with no math content. Intriguingly, the growth was most notable for kids whose parents were anxious about math at the beginning. Those parents often had negative memories and feelings about the subject. The more anxious the parents were, the bigger the improvement in the child's math performance.

The study made plain that parents need to know how to talk about mathematical concepts with young children and they need to believe both in their own ability to do that—whatever their history with math—and in their child's potential to master the subject. "We changed their expectations and values of math for their kids," Talia says. Instead of projecting their own math anxiety onto their children, parents came away with a more realistic sense of their children's capabilities. Providing them with an easy technique for talking about math was like giving them a map—it was one of several way points leading to the North Star.

Stated Beliefs Versus Actual Beliefs

By zooming in on knowledge and beliefs at TMW, I saw the critical importance of what individual parents know and believe. But getting to know the parents we worked with showed me something else: It helped me zoom out and see that our knowledge and beliefs as a society are just as important.

We say parents are children's first teachers and then fail to tell them how to teach.

We say children are our future, but we invest less in children's early years than any other developed country does.

We say we believe in the American Dream, which embraces growth and progress, yet we have set up a society that is so unequal that the dream has become almost impossible to achieve for far too many.

And we say, in the name of American individualism, that parental choice is sacrosanct, yet choice is only a possibility if there are realistic options, and most parents in our society have none.

The disconnect between what we say and what we do, between our stated beliefs and our revealed or actual beliefs, has direct—and sometimes dire—consequences for individual parents. What each mother or father knows and believes about their roles and their children's possibilities is intertwined with what society believes.

Too often, we fail to recognize how parents' daily lives shape the thousands of choices they face as they raise a child, from the relatively mundane, such as what to serve for breakfast, to the more complex, like how to pay for college or whether it's even a possibility for a child to think of going to college. One of the more obvious choices is the question of whether a mother (and more recently a father) stays home with young kids or works outside the home. It's a question that is particularly relevant in the early days of a child's life. Some of us think staying home is essential; others believe it's essential for women to remain in the workforce. I'm not making a value judgment here. I worked during my children's early years, but some of my closest friends and colleagues stayed home. Whatever your opinion, and whatever the ideal may be, when I looked into the statistics, I found that most mothers (about 70 percent) work outside the home. In 2016, only 28 percent of mothers (and 7 percent of fathers) stayed home with children, according to the Pew Research Center.[20] Diving deeper into the numbers, I was struck by the fact that financial pressure works in both directions; it makes it harder for some to stay home and harder for others to stay in the workforce. It is only a minority of parents who truly get to make a choice.[21]

The Choices Don't Add Up

Consider two mothers. Talia, the postdoc who was studying how parents talk about math, now spends her days at home with her two young children because of the exorbitant cost of childcare. Jade, on the other hand, spent twelve years, through her kids' childhoods, as a barista at Starbucks because her family needed the income and benefits. One wanted to work and couldn't afford to. The other wanted to stay home and couldn't afford to. No matter your beliefs, the choices don't add up.

Talia went to Barnard and then worked as a research assistant in an early child development lab at Wesleyan University in Connecticut. She chose to do her graduate work with Susan at the University of Chicago, drawn by the strong Jewish community in the city and by the spirit of the place, in spite of the weather. "We're cold but we're happy," her future colleagues told her. As Talia was growing more confident in herself as a scholar and researcher, she fell in love. Her husband, Justin, started medical school the year after they got married. And she got pregnant with their first child while she was still a graduate student and Justin was still in medical school.

Having a new baby while doing a PhD opened Talia's eyes to the many ways, big and small, that workplaces are not set up to support new moms. She was lucky to have her own office, which gave her somewhere to pump breast milk. But she needed to keep her pumping supplies in the refrigerator, which was in her department's main meeting room. "I would always have to interrupt," she says. Once her daughter started going to preschool, Talia had to leave work at 3:30 to pick her up. Her workplace was flexible, but she knows that not all employers are so kind.

Talia graduated with her PhD while her husband was in his last year of medical school. She decided to continue working with Susan as a postdoctoral fellow, a step many PhDs take in their journey to becoming researchers. Soon, Talia and Justin learned she was pregnant again. They were excited, but now they also had to consider the finances of sending two children to childcare. "Our monthly salaries combined were not enough to cover childcare and rent and food, plus other basic necessities for our family," Talia said. To pay for childcare, they would have to dig into their savings, and that is what they planned to do.

Talia had her baby boy in February 2020 and took three months of maternity leave. When she returned to work, the pandemic had changed everything. Childcare for the baby was closed, as was their daughter's preschool, so both of her children were now home all day. She quickly became overwhelmed. Justin was working long hours as a resident in internal medicine at the hospital while she was trying to balance breastfeeding and nap time between Zoom meetings. Even though Justin's parents live nearby, they couldn't help because Justin's work put him—and anyone around him—at high risk for COVID-19.

"To actually get any of my own work done I would have to work at night, but my son was only three months old then. He wasn't sleeping through the night yet," Talia says, sounding exhausted just at the thought of what she had gone through. If she worked late, she got only four hours of sleep. She felt a deep sense of responsibility to her work. Even with a very supportive environment, Talia recognized that the work had to get done. "I couldn't be upset at Susan asking me to get some data ready for the next meeting," she said. "It's not fair of me to say, 'Keep paying me but I'm not going to do my job.'"

Talia and Justin thought maybe the situation would improve when their daughter went back to school once it reopened. But when they crunched the numbers, it didn't make sense for Talia to work at all. "It was going to end up being more expensive for me to work and to keep her in school." They realized that if Talia stayed home, their daughter could do half days instead of full days at school and there would be no childcare costs for their son. Everything would get a lot cheaper. So, after a BA, a master's, and a PhD in psychology, after professional success that others (and I) could only dream of (a paper in *Science*!), Talia made the difficult decision to stay home. Instead of being an impossible $1,000 in the red every month, they would have an extra $700 to spend on food and other necessities. "It was a no-brainer to say that I should stay home," Talia says. "If childcare didn't cost more than my entire salary, I would be able to work." (Paradoxically, the fact that she quit is what allowed me to hire her as a consultant for a few hours a week to help develop SPEAK. That was an amount of time that was viable for her, and I could afford to be flexible about when she did the work.)

With the numbers in front of her, Talia knew what she needed to do. But psychologically, it was difficult to come to terms with the decision. For one thing, the lab was already understaffed. She didn't want to leave Susan, her boss and academic mentor, in the lurch. "She's a part of my family. I've known Susan longer than I've known my husband." But in a more profound sense, staying home didn't fit with Talia's vision of what she wanted for herself or her family. "This is not where I intended to be," she said. "I gave up any hopes of pursuing the career I want to have for this family that I love. All of a sudden, I became the woman who sacrificed to support her husband's career. It's now my role to not have anything for myself, which is not what I want for myself or to model for my kids."

But it didn't matter. "For me, working was becoming a luxury we couldn't afford."

Jade had the opposite problem. She always wanted to be a stay-at-home mom. Her own mother worked part-time and eventually left the workforce entirely. In her devoutly Christian family, her aunts also stayed home, and so had her grandmother before them. "It's just what I grew up with," she said. "It was the thing to do."

Jade has a direct, no-nonsense demeanor she's inherited from her mother. One of her highest values is being a good Christian, or "showing God's love to everyone." (She admits that desire occasionally conflicts with her instinct to tell it like it is.) Her religion also bred a desire to stay home with her children—not because of anything doctrinal but because of the culture. "A lot of the women in church stay home and they're able to do all those extra things during the daytime to help their kids," she said. "They're able to volunteer. And to be available on Sundays." If you work in a retail job, Jade points out, you may have to work on Sundays.

As a young woman, Jade imagined all the things she would be able to do for her own kids by staying home. Mostly, she just wanted to be there for them, to ensure that all their needs were met, without having to balance home life with the stress and demands of a job. And for the first two years of her son's life, she did stay home. To make it happen, Jade and her husband, Brian, lived with her parents while he worked days and went to school at night to get his teaching degree. But when Brian had to begin student teaching, he had to give up his other job and the health insurance it provided. This meant that Jade needed a job with a decent wage and decent insurance. She found Starbucks. The hours were flexible—she could work early in the morning or late at night, so she didn't miss out on so much valuable time with baby Nathan—and if she worked at least twenty hours a week, her family could have health insurance.

For a time, their new life went like this: Jade worked the first shift at Starbucks, from 5:30 A.M. to 1:30 P.M., and Brian stayed home with Nathan. Jade got home midafternoon and took over while her husband went to work. If their shifts overlapped at all, family members would pick up the slack. It worked but just barely. Jade and Brian hardly ever saw each other.

While she worked pulling shots of espresso and pumping hazelnut syrup into lattes, Jade couldn't stop worrying about what was happening at home: *I wonder what he's doing. Is my baby okay? Does he miss me? If he gets in trouble, are they going to discipline him the way I discipline him?* She felt her mom had always been a bit harsh with her punishments—and Jade hoped she wasn't as firm with Nathan. Emotionally and physically exhausted, sometimes Jade just couldn't stop the tears from flowing.

After a few years, Jade had hope. She excitedly moved her young family to Florida to take a new job with a relative while Brian got a job teaching in a private school. At first, Jade's new job was a great fit. As she was the only one in the office, she'd often bring four-year-old Nathan along since they still couldn't afford any decent childcare, and she also had the freedom to finish her work at home in the afternoon.

But health insurance was still a problem. The new business couldn't provide it, and her husband's school offered a plan that was out of their budget. So in the evenings and on weekends, Jade went back to Starbucks, sometimes working as many as fifty hours in a week in the two jobs combined . . . on top of raising Nathan.

Then Jade got pregnant again. Her plan was that Nathan would start preschool and she would bring the baby to the office. "Newborns—they sleep a lot! And I figured I could probably get a lot done in my four hours there," she said. But when her boss refused to let her bring the baby to the office (I suppose he thought a baby would be too disruptive), Jade was devastated. "He wasn't very supportive of me as a working mother," she said. "I felt slighted because I did a lot of work. I moved my whole family down there. And of course, I would work! I would never not." She tried to argue her case, but it wasn't up for debate. Jade quit. The experience was so upsetting that she got tearful telling me about it, more than ten years later.

Thank goodness for Starbucks. Jade upped her hours and took advantage of the many benefits the company offered to working families: twelve weeks of paid maternity leave, generous health insurance, and a private space in the coffee shop where she could pump. "Even though I was just a barista, I felt like I was heard as a mother." She wished she didn't have to keep working, but if she did, at least it was in a supportive place. Eventually, the family moved back to Illinois, but Jade continued to work for Starbucks for a dozen years in all.

Jade considers herself quite a traditional mother, yet sees the limits of the choices society currently allows. She imagines a world where employers acknowledge how hard it can be to juggle work and family, including during reviews. She would love to see higher salaries and reasonably priced healthcare. She's adamant that she doesn't want "free money." She doesn't want a handout or to be thought of as lazy. She encourages her children to pray and ask God for help, but she wouldn't ask anyone else. Nevertheless, she knows what it's like to feel totally on your own at a time when you most need support.

One Way to Nurture a Brain . . .

There are plenty of ways to successfully raise a child.

There is only one way to nurture a brain.

The comedian and actor Michelle Buteau wrote a hilarious (to me, anyway) essay in *The New York Times* about the shock of discovering that she and her Dutch-born husband had polar opposite ideas about how to safely shepherd their toddler twins through childhood.[22] Their approaches reflected their cultural backgrounds. Buteau was raised in New Jersey by Jamaican and Haitian parents whom she describes as overprotective. In a restaurant, her mother kept Michelle in her lap. ("That was awkward in the teenage years.") But in the Netherlands, her husband grew up in a culture where parents left babies in their strollers on the sidewalk outside restaurants to nap while the adults had a meal inside. "To my husband, I'm not just a helicopter mom. I'm a drone-on-top-of-a-snowplow mom," she wrote. Whereas her message to her casual and relaxed husband is: "Toddler time is not a Jimmy Buffett concert!"

Buteau ends her essay asking: Who's right?

The answer is that there is no one way to raise a child well. Staying home is great. Entrusting children to childcare if it's high quality, or to loving family caregivers, is great, too. But whatever approach parents choose, whoever is providing daily care for a child, there is only one way to nurture their brain: serve-and-return interactions. And it follows that we know very well how *not* to nurture a brain. *Don't* disengage. *Don't* go quiet. *Don't* ignore children (except, perhaps, when they're safely napping outside the bistro).

And what do parents need? They need to know how a healthy brain is built. To know that loving adults are brain architects. To believe that what happens in the earliest years of life plants the seeds of success for a child's future. To have a society that knows the importance of healthy brain development. To have a society that supports parents and caregivers in their critical role as brain architects. And to have a society that believes that brain development is our North Star.

We know the coordinates leading to that North Star. If we follow them, I believe we can get there. But, of course, knowledge and beliefs are not action. What we do matters more than anything else.

SIX

BUILDING FOUNDATIONS AND BUILDING STURDY BOATS

"Scarcity in one walk of life means we have less attention, less mind, in the rest of life."

—SENDHIL MULLAINATHAN AND ELDAR SHAFIR[1]

When I founded the TMW Center, if you had asked me what I hoped for, I would have said Randy was it. He was a father who signed up for our longitudinal study. After working with Randy and his family for a few years, my team and I have gotten to know him well. Watching Randy engage with his kids is a picture-perfect example of how we can take ivory tower science into the real world and show parents how to use the 3Ts to optimize everyday moments and build a child's vocabulary, reading ability, and math and spatial skills.

Randy showed us plenty in return, like how creative parents can be. Between visits, families in our study use a wearable recording device called LENA. Like a Fitbit for conversations, it slips into a child's shirt pocket and captures how much the child is hearing and speaking. To get his son, Julian, to cooperate, Randy said, "I got him thinking it's an Iron Man heart." Julian, who was two at the time, believed his dad. The little boy felt stronger and tougher whenever he put the device on. We've passed the Iron Man trick on to other parents as well. It works!

But Randy showed us something even more important. After participating in our study, Randy knew exactly how to engage with his kids and why it mattered. He came to believe wholeheartedly in his power as a brain architect. The problem was that knowing and believing took him only so far. There were frustrating limits to what he and his family could do on their own, given the economic realities they faced. For me, Randy provided a stark reminder of the societal constraints that keep us from focusing on the North Star of brain development.

Randy first spotted one of our advertisements while riding on a bus along Chicago Avenue. I'm surprised he noticed the poster at all, because he was in the middle of a very long and tiring day. His pickup truck was in the shop—that's why he was on the bus in the first place. After spending hours at one job, he was on his way to a second, where he would log several more hours. Randy was talented across a range of construction jobs: remodeling, painting, plumbing, demolition. Most of the year, however, he maintained parking lots. He tore up old asphalt and laid new, he sealed cracks and painted lines to mark handicapped spots.

Randy routinely worked six days a week, leaving the house at seven in the morning and logging twelve hours a day or more. Sometimes he worked through the night so that the parking lots would be dry come morning. The work left him physically depleted. If it was a 90-degree day (not as rare an occurrence in Chicago as you'd think, given our harsh winters), working for hours on the sizzling black asphalt was like working in an oven.

On the bus, tired though he was, Randy perked up when he saw the bright blue TMW poster announcing that we were looking for families with children thirteen months or older to participate in a program for parents. *Learn the 3Ts*, the ad said. *Tune In, Talk More, and Take Turns. Created for parents just like you who want to learn new and easy-to-use ways to give your young child the best start in life.* Randy felt someone must have been reading his mind. His son, Julian, was just the right age and giving Julian and his sister, Jaylani, who was a year older, the best start in life was exactly what Randy wanted to do.

Randy dreamed of college for his children. That was his slightly hazy vision of what lay on the other side of the river for them. What he knew with clarity was that he wanted more for his kids than he had achieved himself. "I don't want them working my job," Randy said. "I want them to sit in the desk in a nice, air-conditioned room. Have their brains work for them, not their hands."

Already, though, Randy was worried. Jaylani, chatty and affectionate, was thriving in preschool—every day, she came home with new words and there were regular reports on how much progress she was making. But Julian, though he adored his sister and soaked in everything she said, spent much of his time in a different kind of environment. While Randy and his wife, Mayra, were at work, Julian was in a home childcare program five and sometimes six days a week. The people who ran it were kind, and Randy thought they were probably doing their best. But he feared that Julian was missing out on attention, activities . . . he wasn't entirely sure what, but something. Very often when he picked up his son at the childcare, he found the children planted in front of a blaring television while the harried caregivers tended to crying babies. Randy knew there were better places, but this was what they could afford.

When he saw the TMW sign on the bus, Randy picked up the phone. Here, at least, was something he could do.

School-Ready, Future-Ready

Randy saw the link between his children's earliest years and their futures even if he didn't yet know the science behind it. To be ready for college, Julian and Jaylani had to be ready for kindergarten. To be ready for kindergarten, the two children needed strong brain foundations. And those foundations would be built by the brain architects—in this case, Randy himself, Mayra, and the children's other caregivers. The process of building strong brains had begun on the first days of their lives. It wasn't waiting for their first days of school.

That first day of school is significant, however. It is a threshold, marking a child's entry into a new world. That first day can be thrilling or terrifying, full of smiles and tears, or both. I remember my daughter Amelie's first day of preschool especially vividly. She was just three. After watching her older siblings head to school, Amelie could not wait to get her chance to be with the "big kids." We picked out a new outfit—a mint-green shirt featuring a pirate Hello Kitty, pink leggings, and sparkly sneakers. As I walked her the few blocks to school, we were both full of excitement and eagerness. The classroom was bright and welcoming. The teacher, Ms. Abella, greeted Amelie warmly. We found the cubbyhole with her name on it. Many of the other children were the younger siblings of Genevieve's and Asher's classmates, so Amelie already had friends. Laughing and exploring, she pulled me to the shelf of books and then the sand table.

My baby seemed to be completely in her element. She was my third child and I thought I had this, so I was not prepared for what happened next. When it was time for me to go, Amelie began to wail and cling to me. Though I couldn't bear leaving her in that state, the teacher firmly told me to go (and later, that Amelie settled in after I left). Day two was no better. Throughout the first week, each new parting was only marginally less wrenching than the last until she finally, somewhere in the second week, let me go without tears. (The teacher also suggested that Amelie might have a future in theater.)

School requires young children to do many things that are new: sit quietly in a circle, line up, pay attention. Kids encounter letters and numbers and days of the week, new ideas and lots of new people. Emotionally, behaviorally, and academically, some children are ready for this new experience; some are not. At its most holistic—and it should be holistic—the concept of "school readiness" encompasses the full set of skills and knowledge a child needs to function successfully in school. It is usually measured across four domains. One is academic—early language, literacy, and math skills and overall cognitive development. The second and third domains are executive function and social/emotional development, which give kids the ability to wait their turn, use words to express feelings, and persist when frustrated. The last domain is physical health.[2] The first three are tightly connected to foundational brain development. These are the skills that are developed with strong conversational input.

The whole idea of school readiness raises the question of whether children should be ready for school or schools should be ready for children. Ideally, it is both. As the American Academy of Pediatrics says in its policy on readiness, "It is the responsibility of schools to meet the needs of all children at all levels of readiness."[3] But differences in school readiness are also a reflection of the early disparities that have already taken hold. There's no way to sugarcoat the dismal fact that the majority of American children aren't ready for this momentous first day of school, at least according to standard measures. According to a 2020 report from Child Trends, a leading national research organization, more than half aren't "on track" in at least one important area of development that will underpin future success. And children from less advantaged homes are less likely to be ready for school than their more affluent peers.[4] Here in Illinois, for example, only about a quarter of children in the state are truly ready across all four domains, and nearly 40 percent are not ready in any.[5]

Those kids have to play catchup before they've even begun. The truth is, it's difficult to catch up if you start out that far behind. Third grade—or the age of eight—appears to be a critical juncture. If kids arrive in third grade on track, most school districts can keep them on track, but very few can take a group of children who are a year behind at that point and catch them up by the end of high school, according to Elliot Regenstein, an expert on education policy.[6] Those statistics make it clear that if we want to improve the prospects of all students, we have to start even *before* the first day of school.

Mounting evidence shows that the constellation of skills that make up school readiness is predictive of later success. For example, in a large study of more than two thousand ethnically diverse low-income families, a child's early learning environment (including the quality of parent-child interactions) predicted their academic skills in fifth grade.[7] Other research shows that five-year-old children who are ready for school are less likely to drop out of high school as teenagers.[8] The effect reverberates into adulthood. The kids who were school-ready at five have a better chance of reaching the middle class by the time they're forty.[9] They also are likely to have lower rates of chronic disease and substance abuse and to have better health prospects generally.[10] The reason? Skills build on skills, increasing in strength and solidity as they go. Children whose environment helps them to build good brain foundations learn more efficiently as they get older and encounter more vocabulary, more ideas, more sophisticated math concepts.

Out of all the cognitive skills we can help young children develop early on, there's one that stands out as a particularly strong indicator of what's to come: language.[11] Like the headwater of a river, a strong start in language predicts a cascade of skills that flow from it. That means if a parent or caregiver provides a rich language environment full of serve-and-return interactions in a child's first years of life, the child is likely to develop a larger vocabulary and stronger communication skills than a kid who doesn't get the same kind of engagement. And those early vocabulary and verbal skills, in turn, predict an impressive array of achievements, skills, and attributes that will manifest at the age of eight: brain processing speed, vocabulary size, reading achievement, grammatical development, phonological awareness, working memory, and IQ. To put it another way, building strong brain connections in the first years of life is like creating the hard drive of a computer. With good built-in processing speed and memory capacity, you will be able to add software updates indefinitely.

A pile of studies supports the points I've just made, but there's one study in particular that's worth highlighting. In 2018, a group of researchers at the nonprofit LENA showed the long reach of language exposure in the earliest years of life using the recording device and processing software they developed.[12] They had been following a group of several hundred children and families for more than ten years. What was unique about this study was that when the children were babies, the families recorded everything that was said around each child every day for six months, for a full day each month. I do mean everything. These researchers didn't just record an hour of talk a day and extrapolate the total language exposure, as earlier scientists had. Thanks to new technology, the same LENA device we use, they were able to record absolutely everything the children heard. An early study showed that while the overall volume of language—the sheer number of words—that children heard was important, the amount of conversational turn-taking between adult and child was more important for early language development.[13] (To be clear, the study does not prove that one thing causes the other but suggests they are related, and the findings line up exactly with the work of researchers like Rachel Romeo—the speech pathologist and neuroscientist who studied children's brains when they were in a scanner.)

But it is the follow-up study that I think is most remarkable. By the time the children were finishing middle school, it turned out that their early language exposure predicted both their cognitive and their language skills ten years later. On tests of language and IQ, the children who had engaged in more conversational turn-taking with their parents when they were little scored significantly higher. One six-month period in children's early lives proved particularly important: The time between eighteen and twenty-four months of age, when language really takes off and many kids learn multiple new words each day, was most clearly tied to later language and cognitive outcomes and appeared to be a crucial window of opportunity.

This is what's at stake for parents like Randy and kids like Julian. It is why the things individual parents do at home during that critical period will reverberate so widely through their children's lives—and throughout society as a whole. Parent and caregiver talk and interaction is the key to building strong cognitive abilities. That's why it is the driving force behind everything we do at TMW. When parents tune in, paying attention to their child's interests and emotions, they are building their cognitive skills and school readiness. When they share books with their children routinely, engage in conversational storytelling, and elicit responses to words and actions, they are building cognitive skills and school readiness. The more time kids spend doing those kinds of things with their parents, the stronger their cognitive growth. These long-term studies are further proof that no matter what background adults bring to parenting, they can boost their children's brain development by tuning in, talking more, and taking turns.

Time Is of the Essence

Randy turned out to be a natural at the 3Ts. He and Julian met with Michelle Saenz, one of our amazing home visitors. Wearing his ever-present black Bulls ball cap over his shaved head, Randy happily embraced the 3Ts and regularly got down on the floor to play with his son. That wasn't something the men in his family traditionally did. Nor did they do what we call tuning in. Raising four boys, Randy's mother and father were strict and traditionalist. "We didn't talk about thoughts and feelings or anything like that," Randy told us. He grew up thinking that learning happens in school, and it never occurred to him that anything could be done to build babies' brains in the early years. "I always thought the other way, that children were born smart," he said. But he was thrilled to realize that he could help build Julian's brain.

You could say that Randy became a true believer. At a family gathering at his mother's house after he'd been part of TMW for some time, one of Randy's brothers found him playing with Julian. "Man, what are you doing on the floor?" the brother asked. An evangelist for the 3Ts, Randy delivered a monologue about the benefits of getting down to your kid's level and paying attention to their play. "Every little thing counts, everything!" he said.

During his early TMW sessions, Randy would settle in on the soft, gray carpet of his living room floor and playfully scoop Julian up with one arm onto his lap, smoothing his son's black hair and gently tickling his soft belly. Talking to Julian, Randy easily switched back and forth between Spanish and English, a reflection of his mixed Mexican and Irish heritage. When Julian picked up the alphabet blocks, Randy slowly and steadily helped him stack them up, counting as they went ("one . . . two . . . three . . . four . . . five . . .") until they formed a tall and precarious tower in front of the little boy.

"Drop it, drop it." Randy nudged Julian, encouraging him to tip the tower over. Julian gazed up at his dad, his eyes twinkling with delight as Randy added more and more blocks. When the stack, and the counting, reached sixteen, the tower came crashing down.

"Boom!"

Randy especially loved using his passion for Chicago sports teams—the Bears, the Bulls, and the Cubs (never the White Sox!)— to help his son's brain grow through math talk. Whenever they flipped on a baseball game, Randy counted strikes and bases run, he kept score, and he even pointed out the numbers on each player's jersey, which is how Julian eventually began to identify the team members. If the little boy spotted shortstop Javier Báez, he shouted "Nine!"

Randy was making the absolute best of his time with his children. No doubt, he was knocking it out of the ballpark (pun intended). He took our guidance and ran with it. He knew he needed to tune in, talk more, and take turns. He had come to believe in his power as a brain architect. But even though Randy was using the program perfectly, it couldn't offer the one thing he needed more of: time.

Randy's efforts at healthy brain development ran headlong into his economic realities. Whenever he could, Randy tried to get home in the evening in time to spend at least thirty minutes with his kids before bedtime. But to provide for the family, he had to take on whatever extra work he could get. In addition to his regular contracting work and Mayra's job at a dental office, Randy did side jobs—more parking lots, private driveways, drywall—after hours or on his days off to supplement their income.

Randy had always been a hard worker. It was a value his parents had emphasized and from the age of fourteen, he had never been without a job, sometimes more than one. He bagged groceries. He bussed tables. He helped tear down houses on a construction crew.

Straight out of high school, he joined the Navy and spent six years in the service, much of it as a cook. In 2003, his was the second ship to launch missiles against Iraqi President Saddam Hussein during the American invasion of Iraq. (He showed me his medals to prove it!)

Despite his strong work ethic, Randy's lack of higher education left him mired in a job market that failed to provide benefits, a living wage, or regular hours. He wanted nothing more than to emulate his parents, who have held down steady jobs with benefits for their entire working lives. His father, a Vietnam veteran, has worked in the same hardware store for more than three decades. His mother has put in nearly the same amount of time in the locker room of a country club in one of Chicago's northern suburbs. But when Randy and Mayra were starting their family, jobs with benefits and a living wage were hard to find. Instead, he cobbled together as much work as possible. Even though he and Mayra stuck to a carefully calibrated budget, their limited earning power meant money never seemed to go far enough.

Most important, the demands of Randy's work meant giving up time with his family. A half hour a day with his children was so little. Too often, he missed even that because he felt he had to take whatever job came along, even if it meant a late night. On a Saturday, Randy would rather be home watching a baseball game and using math talk with Julian and Jaylani than laying hot asphalt. The loss of Randy's time with his kids was compounded by the fact that the childcare program they could afford for Julian was not of high enough quality to complement what Randy and Mayra were doing at home. The financial stress and exhaustion from work were getting to Randy. The area they lived in added to his concerns. He had grown up in the same neighborhood, and the house was passed down to him from an uncle. Outside, lots stood vacant. Fast-food wrappers littered the ground, and gunshots regularly punctured the night's quiet. Two of Randy's brothers were shot when they were younger. "I tell the kids it's fireworks," Randy explained, "but pretty soon they are going to realize it isn't." He tried to make a difference by volunteering with the Urban Warriors, a youth safety and violence-prevention program run by the YMCA. But at night, he often lay awake feeling guilty, and then he worried about worrying. Would his state of mind affect his kids?

A Sturdy Boat

Parents like Randy want nothing more than to do the work to help their children succeed. But without time and bandwidth and money, it's as if Randy was trying to ferry his son and daughter through childhood in a leaky rowboat. He was pulling as hard as he could on the oars, nurturing them to the best of his ability, but the boat kept taking on water. It was a constant battle just to stay afloat. Raising a family in the United States is expensive. To get a child through high school, a household can expect to spend more than $200,000, or nearly $13,000 per year.[14] Parents need sturdy boats if they're going to get their children all the way across the river and ready to take on college or whatever other kind of opportunity lies on the distant shore. If I may carry the metaphor just a little further, I'd say it doesn't have to be a yacht, just a seaworthy vessel, one that can weather the storms of life without sinking. That boat would be built of wages a family can live on, reasonable work hours, affordable high-quality childcare. In other words, reliable support that is truly . . . supportive.

Over the last fifty years, however, the economic possibilities for someone like Randy have only narrowed. It used to be that someone who did physical labor could earn enough to comfortably raise a family. That is no longer true. Employment and wages are rising mainly in jobs that require higher-level social or analytical skills and that call for greater education or training. Manufacturing employment has declined by about a third just since the 1990s, while employment in knowledge-intensive and service sectors has about doubled. (Hence the growing sense that "intensive parenting" is the only way to help kids get where they need to go.) Wages have been mostly stagnant since the 1970s. What wage gains there have been have gone mainly to the highest-paid tier of workers. As a result, income inequality has expanded alarmingly.[15]

Those facts and figures may sound familiar. We have been hearing about income inequality for years. But here's what it means in practice. Even workers like Talia, with so-called good jobs (jobs with a salary, benefits, and regular hours), struggle if they are at the lower end of the wage scale. A 2020 study that zeroed in on such workers at a major Pittsburgh hospital (they were secretaries, medical technicians, and the like) found that the majority had to resort to extreme strategies such as payday loans or delaying the payment of utility bills in order to put food on the table. They relied on family members for childcare and they often went without some essentials (for instance, health insurance for adults, before the Affordable Care Act, was often seen as an unaffordable luxury).[16]

Then there's the gig economy. Driven largely by technological changes, more and more industries have been disrupted and a larger share of workers engage in short-term or freelance work for all or some of their income. My grandfather was a salaried truck driver for Giant Eagle grocery stores. He owned his home and kept his family firmly in the middle class. Now that kind of job is often part of the gig economy. It's not so easy to measure the rate of participation in the gig economy, but recent estimates are that more than a third of American workers, like Randy, take part in the gig economy.[17] For many, the gig economy means a side hustle that provides some extra cash. That's certainly a benefit that's trumpeted for some stay-at-home parents.

But for others, it's their primary way of earning a living. The advantages are flexibility and (sometimes) low barriers to entry. The disadvantages are a lack of benefits, security, and control over their time, all of which are pretty important if you have a family to raise. Contrary to popular belief, the gig economy isn't composed of just Uber and Lyft drivers or Airbnb landlords; it includes independent contractors, on-call workers, seasonal workers (such as for Amazon during the holiday rush), and many others. According to one report, anywhere from 40 to 60 percent of gig workers (depending on whether the work is primary or supplemental income) say they would have trouble handling an unexpected expense of $400— which means they are living perilously close to the financial edge.[18]

These economic changes and challenges make day-to-day life difficult in many ways, and in particular they have a knock-on, or secondary, effect on parenting. The realities of today's job market are in direct opposition to our growing knowledge of what parents and families need for optimal brain development. Parents who feel financial stress are less likely to feel supported as parents, less likely to set limits or be satisfied with their parenting, and less likely to communicate well with their children.[19] The reality is that economic and other stresses play a significant role in the overall functioning of families. But the link between workforce issues and the developmental issues of early childhood is not something that gets a lot of attention.

Searching for Competence and Conversation

Parents who work outside the home know all too well how enmeshed their jobs and family life truly are. And they know that childcare is the linchpin that can hold their lives together. (It took a pandemic, which closed the schools and the childcare facilities and left parents to fill in, to make everyone else see this.) So just as parents are brain architects for their young children, so are the early childcare providers with whom kids spend much of their time. Yet only 10 percent of childcare settings are rated very high quality, meaning as places that base the rhythm of their days, their activities, their caregiving on what we know about strengthening young children's development. The other 90 percent vary dramatically, and too many are simply custodial. Some 20 percent of children are "in language isolation," experiencing fewer than five conversational turns per hour.[20] (Forty turns per hour should be the benchmark.) I know many of the staff in childcare centers are doing their best, but like Mariah and the dames of the colonial era, who could barely make ends meet working in childcare, most are underpaid, undertrained, and overstretched.

One of the things that suffers most in lower-quality care settings is rich language engagement. The same researchers who recorded so much of young children's early lives found that kids experience vastly more conversational turn-taking at home than in childcare settings—73 percent more![21] That key window for early language input they identified between eighteen and twenty-four months? That's exactly when interactions in childcare settings were at their lowest. Notably, preliminary results from a study of 1,742 infants and toddlers across thirteen countries showed that between March and September 2020, the vocabularies of children who were home with their parents because of the pandemic grew more than would have been expected if they hadn't been home—the study found that those children had less passive screen time or their caregivers read more to them, and it represents one of very few bits of good news that came out of the pandemic.[22] (Of course, this was in the first six months of the pandemic, and as the crisis wore on and parents were stretched ever thinner, this positive effect may well have faded.) "Either caregivers were more aware of their child's development or vocabulary development benefitted from intense caregiver-child interaction during lockdown," the researchers concluded. It's not necessarily all that surprising that there's more talk at home than in childcare. No one is more interested or invested in a child than his or her parent. And at home, there are fewer children vying for attention and probably fewer conversational interruptions. But the extent of the difference and the significance of the window of opportunity for children's cognitive development highlights the critical importance of quality childcare.

With the stakes so high, parents up and down the economic ladder—the vast majority of whom must work outside the home—become consumed with the effort first to find good childcare and then to pay for it. Gabrielle certainly was. An audiologist and hearing research scientist in Omaha, Nebraska, Gabby started looking for a childcare center as soon as she found out she was pregnant. In truth, she had started planning well before that. She had to. Her job offered "pretty dismal" maternity leave. "You either take twelve weeks unpaid, or you use up whatever vacation time and sick time that you have accrued," Gabby explained.

In order to eventually be able to take a maternity leave, Gabby took no vacations for her first few years at her job. After several expensive fertility treatments, she and her wife, Kaila, welcomed their son on November 5, 2019. "He is my little IVF miracle," Gabby told me. They called him Greyson Kennedy and I love the reasons why. Their son's first name evokes Mount Greylock, the highest point in Massachusetts's Berkshire Mountains, where they both grew up, and the place where Kaila proposed to Gabby. Kennedy, his middle name, honors former Supreme Court Justice Anthony Kennedy, who was the swing vote and author of the majority opinion that federally legalized marriage equality in June 2015. The couple even read the final paragraph of Kennedy's opinion in *Obergefell v. Hodges* at their wedding.

Greyson's name is poetic, but his birth was difficult, and he was born with a microform cleft lip. Gabby handled those challenges well. "I wanted him for so long and worked so hard to bring him here," she said. "It's hard to identify what's hard when most days I wake up and count my lucky stars that I get to be his mom . . . It's a crazy love."

When Gabby and Kaila began touring childcare centers, they had horror stories on their minds. A friend sent her baby to a childcare in town and twice was handed the *wrong* child at the end of the day! One place they called said they preferred not to have prospective parents drop in. "That was a bit of a red flag for me." Some places were alarmingly unclean or reeked of diapers. "Kids are messy, and we didn't expect a center that was glistening and brand-new, but some were really, just, dirty." Others were incredibly noisy or featured televisions on carts. "I hate mom-shaming, but I don't think screen time offers any learning, at least [not] under eighteen months or two years." (Her view aligns with the science as well as with the recommendations of the American Academy of Pediatrics.) For a time, it seemed too much to hope that they would find a place where Greyson would get lots of tummy time (they didn't want him in a bouncer all day) or where he could build strong relationships with his caregivers. (The latter would require low teacher turnover, which is especially hard to come by, given that studies estimate annual turnover rates somewhere between 26 and 40 percent in licensed childcare facilities, and, of course, the burdens of COVID-19 exacerbated the problem.)[23]

More than anything, Gabby and Kaila wanted a place where the caregivers talked to the children and engaged with them. They knew that talk and interaction are critical for cognitive development. They wanted a place with an educational approach, that was focused on child development and tracked milestones. Since she works with children who have hearing loss, Gabby knows a lot about the importance of surrounding children in their first three years of life with a language-rich environment. (You can tell we come from similar fields.) She watched the caregivers intently: *Were they talking to the youngest children? Were they practicing conversational turns?*

Finally, on a tour, Gabby walked into the infant room of a center and saw a teacher holding a cooing, babbling baby who was about five months old. "The teacher was looking at the baby with this big smile and these vivid expressions, great eye contact, and responding to each coo and babble with an 'Oh, yeah? Tell me more about that. Ooh, really? No way! That sounds amazing.'" Gabby liked that. (If we had the powers of neuroscientists to look inside the brains of that baby and caregiver at that moment, I'm sure we would have seen the synchronization of their brain waves at work.) To add to the appeal, the administrators at the facility told Gabby and Kaila that all the teachers knew basic baby signs and would work on those with the kids. They sent home materials to encourage families to practice signing, too. "That really suggested to me that they understood the importance of early language learning in any form," Gabby said. As the tour continued, she saw children in classrooms having circle story time, doing art projects, and more. Babies in the infant rooms rarely cried, and when they did, they were immediately attended to.

Gabby and Kaila were sold, but they still had to get Greyson in. They signed up on the waitlist when Gabby was just twelve weeks pregnant, but they didn't get approved for a spot in a center until right before her maternity leave was up. That was one entire year later! "I just obsessively called, and I honestly don't think he would have gotten off this waitlist, but they ended up turning a toddler room into an infant room because of demand," Gabby says. "So they spontaneously had an infant spot the day that I was calling and being annoying. And I was somewhere in the right spot on the waitlist, and it all worked out."

To pay for Greyson's childcare, Gabby has taken three adjunct teaching jobs online at two universities in addition to her full-time job. "One of the classes, the compensation depends on how many students are enrolled. So, I hold my breath," Gabby said. If thirty-two students sign up, she gets the next bump on the pay scale, which covers another month of childcare. "That's literally how my brain works. I'm like, what's another month of daycare? We're a two-parent working family with good jobs. It's still outrageous."

With Gabby's extra work, she and Kaila can swing the cost, but they have to be more conscientious about their budget than they wish they had to be. She and Kaila would like to have more children. "Honestly, I'd have like ten children if it wasn't so expensive," Gabby joked. "We have talked about expanding our family and the limiting factor is finances."

The exorbitant cost is all the more frustrating to them because they know how poorly the teachers are paid. In part to compensate, Gabby gets the teachers gifts for every holiday, big and small, to show how much she appreciates them. "They are the people who are raising our child, basically. In some ways, I wish we had a nanny, because then at least I would know that the money is all going to them. And I know it's a business, there's overhead, they have to pay for the building. I just, in some ways, as expensive as it is, I wish I could pay three times as much so the teachers could get paid more, because I think they deserve it."

Gabby's gift-giving might be a bit excessive (Kaila occasionally thinks so). But Gabby knows she isn't just trying to help the teachers. There's a deeper motivation. Her parents divorced when she was seven and her dad died soon after. Then her mom struggled with mental illness and addiction, leading Gabby to spend time in foster care. "I didn't have the mom who came to your class and talked about her job, or who sent you to school with cute little Valentines for your friends. And so, in some ways, I feel like I'm living this dream of maybe the life I wish I had had," Gabby said. "I'm just trying to be the best parent I can be and sometimes that means I'm a little over the top."

Linked Lives

The lives of children and parents are intertwined. Those bonds are strong and closely tied, established through love and through time as surely as talk establishes durable connections in children's brains. The experiences we have in childhood, as mediated by our parents, will be reflected in much of what happens to us years later. Gabby's life was shaped by her parents' woes. Greyson's life will be shaped by Gabby and Kaila's choices. Randy's life is shaped by his parents' work ethic, and by his pride in his military service. And Julian and Jaylani's lives will be shaped by Randy's decision to call TMW.

But the way people's lives are intertwined goes beyond parent-child relations—far beyond. When I think about Gabby's gift-giving or Greyson's name, when I think about Randy's work with young boys in the Urban Warriors or the way he talked to his brother about playing with Julian on the floor, I am reminded of how tightly we are all linked to one another, often in ways that we could never have expected. And the links spread even further. The policies of the country we live in weave their way into our lives and can have both immediate and far-reaching implications. (Consider the bond between a Supreme Court justice and little Greyson Kennedy.) Yet too often, our persistent focus on ourselves as individuals blinds us to the ways in which we are interdependent and always have been, and our focus on taking responsibility for ourselves blinds us to how dependent we are on certain kinds of institutional, societal support. After all, the founding fathers recognized that democracy required common schools so as to create a citizenry educated enough to be informed voters. Today, we must recognize that the fates of parents and children are bound up in the fate of societies, of nations.

In countries where there are more family-friendly policies, where there is universal childcare, for instance, or paid family leave, there are smaller language gaps between rich and poor children than there are in the United States, according to the Wordbank Project that Michael Frank runs at Stanford University.[24] Since strong cognitive development flows from strong early language development, those smaller language gaps benefit all of society in the form of a more educated workforce and other positive public outcomes.

That means the challenges that parents face affect us all. How can parents be expected to build their children's brains if survival demands they are rarely at home during their children's waking hours? Scores of parents work hard to keep their families stable, often holding down more than one job, including jobs that we regard as "good jobs," like Gabby's, and like Talia's, yet they still either lack the ability to pay for childcare or can just barely afford it. "We've created a context in which parenting is virtually impossible," sociologist Jennifer Glass of the University of Texas at Austin told me. "No one can afford to have a child unless they're millionaires, without marked financial sacrifices that will escalate over time. Having two or three can literally bankrupt you." Jennifer studies family dynamics and parenting. She has always been struck by our romantic ideal that parenting is a joy not to be missed, that it's always a blessing to have children. But if you look closer, you find that most parents are not as happy as you'd expect. That doesn't mean they don't love and find joy in their kids. It means that their boats are leaking, and they are struggling to ferry their children across the river to a bright future.

With her colleagues Robin Simon and Matthew Andersson, Jennifer did an eye-opening study that asked whether this was an American phenomenon or a global one.[25] They looked at parent happiness across a range of twenty mostly Western developed countries (because that's where they could find comparable data) and created a database of policies that assist parents (things like family leave, flexible work schedules or control over schedules, and combined paid sick and vacation leave). Then they compared parental happiness in countries with family-friendly policies and those without them.

"We were just blown away," says Jennifer. Everywhere they looked, parents were less happy than non-parents. But in countries with more generous family policies, particularly paid time off and child-care subsidies, the disparities in happiness tended to be smaller between parents and non-parents. And importantly, in those countries where family-friendly policies boosted the happiness of parents, the policies did *not* reduce the happiness of non-parents (for instance, by putting more pressures on them in the workplace).

Randy was certainly one of those unhappy American parents, working long hours to make ends meet and lying awake at night filled with worry about his kids' future. Fortunately, a few years after we met him, things changed for the better for him. Feeling he couldn't get through another hot summer of parking-lot work, he started searching online for job listings and found his dream job: doing maintenance at a YMCA in a nearby suburb. It offered a generous salary and benefits. To go to the interview, Randy took the day off, telling his boss he had to take his kids to the dentist. The boss docked him $100—a normal practice for any day off. Randy was so nervous he sweated through the interview. When he found out he had gotten the job, he was overjoyed.

A weight has lifted from Randy's shoulders. He has an easy ten-minute commute instead of traveling all over Chicago's suburbs and into Indiana. Instead of working on hot asphalt all day, he is in an air-conditioned building. After one year on the job, he will be able to use vacation days (without losing pay) and he has access to a retirement savings plan.

Most important, the new job has given Randy time to spend with his family. Both kids are now in school full-time, but he is regularly home for dinner. He always has Saturdays and Sundays off, so he and Mayra can make real weekend plans with the kids. On Saturdays when Mayra has to work, they no longer dip into their limited funds for a babysitter. Randy is there.

One of the things Randy is most looking forward to doing with his new paycheck is taking Jaylani and Julian to their first Cubs game. "When you see the green grass at Wrigley Field, you'll never forget it," he says.

I think the kids are even more likely to remember all the time they are spending with their father these days and all the conversations they are having. From their sturdier new boat, the outline of the far shore just came more clearly into view for Randy. What I see are the cascading cognitive skills that will strengthen Julian and Jaylani's futures.

SEVEN

MAKING MAPS AND NAVIGATING THE TORRENT

"Sometimes in our lives, we all have pain, we all have sorrow . . ."

— BILL WITHERS[1]

Now that my kids are grown, I depend on my friends to provide my "cute kid" fix. A dear friend recently sent me a video of her son Jesse. A four-year-old with rosy cheeks and a soft swish of brown hair across his forehead, he sits in front of a white table. On the table is a marshmallow. Staring at it, Jesse mumbles something quickly.

"I can't understand what you're saying," his mom, Katie, says.

"The marshmallow's getting tinier," Jesse says earnestly as he looks down at the undersize marshmallow, "because it's been sitting out *longer*." Of course, the marshmallow hasn't changed, but studying its shape seems to be helping Jesse avoid devouring it. And that is the goal here. Katie has re-created the famous marshmallow experiment for her son.

In the early 1970s, researchers at Stanford University presented preschoolers with a treat—a marshmallow or a pretzel, depending on the child's preference—and said they could eat that one immediately or have two later if they could wait fifteen minutes while the adults were out of the room.[2] Then they left and let the video cameras start rolling. The kids tried everything from singing to themselves to putting their hands over their eyes to avoid giving in to temptation. One even prayed to the ceiling. Years later, the researchers reported that the children who managed to wait were more competent as adolescents than those who didn't wait. They also had better SAT scores and better physical health.[3] The ability to delay gratification seemed to be the secret to success.

In my friend's video, Jesse is trying hard to wait. He knows he will get an extra marshmallow if he can just hold out. He stares at the marshmallow a few seconds longer until at long last an alarm sounds from Katie's cell phone. He has done it. Jesse gasps and grins before jumping up from his seat and shouting, "YES!!!!!" Then he does a victory dance. His mother beams with pride.

Katie is far from the only parent training her kid to delay gratification. The #patiencechallenge recently went viral on TikTok. Celebrities like Kylie Jenner and Gabrielle Union-Wade filmed their young children resisting temptation or failing to. Kylie put a bowl of M&Ms in front of her three-year-old daughter, Stormi, and promised three of the candies if the little girl could wait until Kylie went to the bathroom. For fifty seconds, Stormi stares at the bowl. At one point, it looks like she might break. She slowly and hesitantly reaches out for the bowl. But then she begins to sing a song that goes like this: "patience . . . patience . . . patience." With each repetition of "patience," she seems to resist the pull of the candy. Gabrielle's daughter, Kaavi, couldn't hold out. She gobbled up the candy right away. To be fair, she was quite a bit younger—still in diapers—and probably wasn't really expected to be able to wait.

Patience or no patience, it is all awfully cute. But such videos, whether from my friend Katie or celebrities, aren't just fun. They are a sign that the idea that parents can help their kids learn to control their behavior has gone thoroughly mainstream. What those parents are really developing in their children is executive function, the skill employed when we control our impulses.

Clever though it was, the original marshmallow study was flawed. Among other problems, it failed to take account of other differences among the children that would have affected their outcomes.[4] But the study caught the public's attention—and has stayed there—for a good and important reason. Executive function has been repeatedly found to be critically important in school and beyond.[5] Intuitively, parents understand that the skills kids deploy to resist a gooey marshmallow are the same skills they need to avoid having a meltdown waiting in line at the supermarket or to sit still in school—skills that can feel frustratingly hard to achieve. Often, during the third or fourth session of TMW's home visiting program, a parent will say something like this: "I love that my words are building Jimmy's brain. Can I also use them to get him to finally start behaving?"

The short answer is yes. I didn't realize it when I started on my journey outside of the operating room. I was focused on language input and parallel language development, not executive function. Like many people, I assumed that executive function was unrelated, that it was innate. As I dug deeper into the scientific literature on how the brain develops in the first years of life, I learned otherwise. Executive function is a muscle that parents can help their children strengthen. The same quintessential serve-and-return interaction responsible for building a child's language ability is also at the heart of developing children's ability to regulate their emotions and behavior in the early years.[6] That means the very thing we were focused on at TMW—language input in the first three years of life—turned out to be key for laying the foundations for strong executive function as well. As soon as we appreciated that, we began adding lesson modules on executive function and behavior to the repertoire we shared with parents and teachers.

That said, executive function does take time to develop. The results of parents' efforts won't begin to show up until kids are well past their third birthdays, and the skill will continue to develop throughout childhood and adolescence. Even my college-age kids are still working on it! Many parents underestimate how long it takes. When TMW investigated parents' beliefs about what children are capable of between zero and three years, more than 70 percent thought that children two years and younger could share and take turns, and more than half thought they should be able to control their emotions and impulses.[7] Neither skill emerges until age three or four.

Getting to know a host of TMW families highlighted for me the fact that time is not the only consideration. Developing executive function requires a calm, caring, supportive environment, or something approaching it. Anxiety, harsh language, and violence all detract from the important process of developing the brain circuitry that underpins executive function. That circuitry is exquisitely sensitive to toxic stress. What happens to us at particular points in life, especially in the critical early years, sets up the body's stress responses for the rest of our lives. It is a cumulative process of development, but there are long-term consequences for early disruptions.[8]

The river of life carries challenges for all of us. But the water is more turbulent for some than for others. For too many families, it becomes the torrent I saw in my dream. That torrent—in the form of racism, poverty, violence—can threaten to sink us. And it can carry (or cause) internal challenges—depression, substance abuse, disease—that threaten to suck us under like a powerful riptide. When society is structured in the right way, its safety nets can protect us from the storm. But society does not provide equally effective safety nets for each of us. The people most vulnerable to societal ills are also the least likely to have a sturdy boat. When parenting means crossing a raging torrent, it can be difficult to escape the dangerous currents.

Laying the Groundwork for Grit

One of the parents who showed me this most vividly was Sabrina. We met her as part of our home visiting program. She and her two-year-old son, Nakai, looked uncannily alike, right down to their rectangular eyeglasses, and Sabrina's sense of wonderment in her son was obvious. "I love seeing his face when he just lights up," she told us. "He's a bundle of fun, he really is." Like so many other mothers and fathers, she was working on getting her son to behave. He could be mischievous. Nakai loved to take Sabrina's DVDs out of their cases and rearrange them. (On the plus side, when his mom handed him a grocery bag, he knew that was his signal to help clean up.) Another favorite game was trying to flip off the bed.

"He is a daredevil!" Sabrina said. "He has no fear, none."

When Nakai got ready to leap, she would tell him. "Baby, we cannot flip off the bed, no." And then she explained that flipping off the bed is likely to end with a big bump on the head or worse.

Sabrina was thrilled to learn that her parenting impulses—the way she talked to Nakai in those instances—were not just preventing a bump on the head but also contributing to her son's healthy brain development. The 3Ts build executive function in much the same way they build more traditional academic skills. By explaining the reasons for her rules, Sabrina was strengthening the connections Nakai would make between actions and consequences. Rewarding good behavior was another good idea. Whenever she "caught" Nakai being good, Sabrina liked to deliver a high five, affectionately slapping his little hand with her much bigger one, something that made the toddler grin from ear to ear. Whatever he was doing right made for a great topic of conversation, a way to talk more.

She had also consciously instituted some positive routines with Nakai and his older brother, including going to the playground every day at ten o'clock, making sure they washed their hands before eating, and bathing regularly. Little things but important things. One of Sabrina's goals while working with TMW was to add even more routine to her children's lives. "That just makes them feel more safe and comfortable," she said. She is exactly right. Scientific literature strongly supports the importance of stability and predictable routines as critical factors in the development of a child's self-regulation and socioemotional skills.[9] The 3Ts are designed to facilitate such predictability. A positive tone and the responsiveness of tuning in are especially important. When parents embed the 3Ts in daily routines, such as bedtime and playtime and eating lunch, the predictability of those routines is key to supporting a child's self-regulation and executive function. And the predictability of routine can do even more. It can build a child's resilience.

The analogy we use at TMW is to think of the behavior-controlling part of the brain as a stoplight with a green signal and a red signal. Green tells us to go, go, go and act on our wants and impulses. Red tells us to stop and think to control those impulses. As adults we use our stoplights all the time to make smart choices. (Or we try to . . .) Our brains developed that control over time, and young children's brains are just at the beginning of the process. As it matures and becomes stronger, a child's brain will develop the power to activate that red light, giving the child the ability to control their emotions and behavior. Until kids can do that for themselves, they need help from the adults in their lives.

Executive function goes beyond allowing children to inhibit their impulses. It helps them map out what they need to do. It includes skills they will need in school, like working memory, which gives them the ability to update the information they hold in mind. (*The teacher said to sort the blocks by color.*) Executive function allows kids to switch from one task to another, an example of cognitive flexibility. (*Now he wants us to sort the blocks by shape.*) It helps them finish putting the blocks away when asked, without getting distracted. (*Now I'm supposed to stop playing with the blocks and put them back in the box.*)[10] Without optimal executive function, children struggle to adjust to school, to focus, and to absorb what is being taught to them. They have difficulty staying organized and planning for short-term projects and long-term possibilities. Worse, research suggests that early emotional and behavioral problems often persist and may even increase as children grow, whereas preschoolers with advanced executive function have better social understanding and go on to superior academic ability and fewer behavioral problems.

As evidence piled up over the last twenty years, it began to seem that executive function (aka grit or noncognitive skills) was the most important capacity of all for academic and lifelong success, even more than cognitive skills like letter knowledge and early numeracy. Books about grit and resilience as the key to success became bestsellers. Kindergarten teachers overwhelmingly picked noncognitive skills like executive function as most essential for school readiness.[11] That's because in a classroom, children's behavior affects everything: how much the group can accomplish, how much fun they can have together, and ultimately how much the children can learn. Of course, as the school readiness research makes clear, cognitive and noncognitive skills are both essential. Moreover, they are complementary; the stronger each is, the better they work together to help children learn. And what we now know—as the #patience-challenge videos show us—is that parents play a powerful role in developing both.

The term "grit" is a reminder that executive function skills are sometimes known as character skills. I've always found the term "character" jarring in this context because it suggests someone's core identity; it implies an inborn trait. Thinking of impulse control that way is as problematic as the idea that young children should pull themselves up by their bootstraps. As Kim Noble reminded us, babies don't have bootstraps. They aren't born with impulse control either. Executive function is the product of a dance between nature and nurture. The genes parents hand down provide a starter kit, but how the skill will ultimately be built depends on the experiences children have. Like reading or subtraction, executive function is a skill that has to be nurtured. It is developed over time in specific parts of the brain, primarily the prefrontal cortex, the last brain area to fully mature.

The special challenge of executive function is that during that journey to maturity, the prefrontal cortex is strongly influenced by and reactive to chronic stress and anxiety. What does this mean? That children's self-regulation develops in response to what's going on in their homes, in their neighborhoods, and in their schools.[12] Many of these problems are systemic. For too many children, those homes carry toxins like lead paint, the neighborhoods struggle with violence like Randy's did, and the schools are overcrowded and under-resourced like Hazim Hardeman's neighborhood school was. Some families must also contend with gangs as well as the scourge of addiction. (Drugs are an enormous problem in rural America, which has been hit hard by the opioid crisis.) I was stunned to learn that the presence of children is one of the strongest risk factors for eviction, according to the research of sociologist Matthew Desmond, author of *Evicted*.[13] Food insecurity is common among the poor. Youthful actions, like an arrest for marijuana possession, are hard to erase and can limit housing and employment options, depending on the socioeconomic status of the person involved. The lives of TMW participants Michael and Keyonna were turned upside down when Michael, a victim of racist policing and a glacial justice system, spent years in jail for a crime he didn't commit. All of it can add up to what researchers call toxic stress.

Of course, you don't have to live in a violent neighborhood or experience racism or be poverty-stricken to face devastating circumstances. After Don died, I worried terribly not just about my children's physical health but about their emotional health. Amelie, who was only seven, became obsessed with death, terrified that I would leave her, too. For much of the first year after her father died, the only way she could fall asleep was if I was lying with her in the bed, her small hand touching my arm or my back for reassurance that I was still there.

On the list of "adverse childhood experiences," now referred to by the acronym ACEs, losing a parent ranks right up there for negative impact. Fortunately, it is rare. But only a very lucky few sail through life untouched. Eventually, the wind whips up and waves form in the river as a storm rolls in. In fact, American children are likely to experience some kind of negative event while they are young—from economic hardship or living with divorce, to abuse and neglect or living with an adult who is mentally ill or addicted to drugs or alcohol. According to the Centers for Disease Control, roughly 60 percent of American adults suffered at least one ACE as a child and one adult in six experienced four or more.[14] These are not just short-term challenges to be overcome. They can have long-term toxic effects. Adversity faced in childhood burrows into the brain and can—if not addressed—disrupt the stability of the circuitry that is built and, therefore, the stability of the child's future. Accumulation of these negative experiences is especially harmful. The work of pediatrician Nadine Burke Harris has powerfully highlighted that having four or more adverse experiences while young puts a person at especially high risk for negative physical and mental health outcomes such as heart disease and depression in years to come.[15] And their negative effects can pass from one generation to the next.[16] If mothers experience stress while pregnant, that stress can affect which genes are expressed in their children once they are born—it can change their genetic programming, in other words.[17] The more stressed parents are, the more likely young children are to have mental health issues such as anxiety and depression. Everything I've just described is entirely out of a child's control, and often out of a family's control as well.

Caught in the Rapids

As crushing as the loss of a parent is, my children were protected in other ways. I still had a good job as a doctor and a roof over our heads. Our community bathed us in support, feeding us for months, and holding us up when we thought we would fall. When life spiraled out of control for Sabrina and her family, they were not as fortunate. The safety nets that might have caught them failed. Those TMW sessions where we talked about not jumping off the bed? They took place in a homeless shelter, a place that Sabrina would never have believed she'd be calling "home."

Just two years earlier, she had been celebrating a promotion at Community Care, a home-based services provider for the elderly. Starting out as a home aide, Sabrina quickly rose in the ranks and became an administrator in the company's downtown Chicago office. (She also has a passion for writing and hopes to be published one day.) Sabrina's husband, Wayne, worked as a laundromat attendant. At the time, they had only one son, Jason, who was eleven. When Sabrina moved from an hourly wage to a salary, it meant their family of three could finally move into the ranks of the middle class.

But their security unraveled quickly when Wayne was diagnosed with diabetes and spent a week in the hospital. Sabrina had been enduring a long commute—getting up at 5:00 A.M., leaving home in the dark, and returning around 8:00 P.M. Wayne had been working nights, getting Jason up for school, making his meals, keeping the house tidy. But in the weeks after his diagnosis, Wayne had trouble regulating his blood sugar, and his poor health limited his ability to care for their son or himself while Sabrina was at work. The strain was getting to all of them, and Sabrina thought she could help get Wayne's health stabilized if she spent more time at home. She first asked to change her hours and then for family medical leave. Her boss denied both requests.

Putting her family first, Sabrina walked away from the job. Just weeks later, she found out she was pregnant with Nakai—a shock since she had long ago given up her dream of a second child. With the extra cost of a new baby to plan for, Wayne's income alone was not enough. Soon they had exhausted their savings and could no longer make rent. They spent a few months at Sabrina's mother's house, during which time she gave birth to Nakai, but the stay was fraught with tension and they had to leave.

In 2019, the average stay in a homeless shelter in Chicago was less than four months and that's about what Sabrina expected.[18] Instead, the family was stuck there for two and a half years. Nakai turned one, then two, in the shelter. Her older son, Jason, turned twelve, then thirteen.

Outside the calm and order of their tiny room, which Sabrina and Wayne had scrubbed clean when they first arrived, chaos and filth reigned supreme. Sabrina described living in the homeless shelter as the worst and most traumatic experience of her life. Immediately, there were warnings about sex offenders in the building. Children, she was told, must be kept close. As a result, the family carried little Nakai everywhere. During his first two years of life, Sabrina made sure his feet almost never hit the ground. And she never let Jason and Nakai use the (shared) bathroom down the hall by themselves. Although she knew that television wasn't great for a very young child, sometimes she had to turn it on—loud—to drown out the high-volume arguments of the next-door neighbors.

Despite the instability and stress swirling around her, or perhaps because of it, Sabrina signed up to be a part of TMW. She wanted to do all she could to give Nakai the best chance to succeed. Rearing children successfully is not easy in the best of circumstances. But rearing them in destructive environments can be almost impossible.

One of the things we talk about during TMW home visits is that it isn't only the number or quality of words that matter, it's the way those words are spoken. The tone and mood of the language is crucial because noncognitive skills are especially affected by parenting style. Harsh reprimands and directives may solve the problem in the short term and may even add up to many words. What they don't do is help build executive function. They undermine it and set the stage for poor self-regulation and executive function in the long run. The best outcomes result from supporting children's autonomy and from a style that combines effective discipline with parental warmth.

But Sabrina's fear for her boys and her determination to protect them from the world around them was so strong that if Jason and Nakai were more than a few feet away from her in the shelter, Sabrina would yell at them. Her terrified tone yanked them back to her side like a leash. After learning about executive function and how it develops, Sabrina wanted to know, What should she have been doing?

To be honest, our team didn't have a satisfactory answer. Optimal approaches to raising a child in a homeless shelter don't exist in parenting books and are often left out of parenting educational programs. How does one balance the facts that the brain needs one thing and life in a dangerous setting may require something else? Our strategies work fine for everyday frustrations—like her son's

determination to flip off the bed—and for those occasional moments when a parent sees their child chasing a ball across the street
as an oncoming car comes dangerously close. What do they do? Yell,
of course! And they should. But what happens in situations like Sabrina's, where emergencies are an everyday occurrence and drama
is part of the daily routine? I think we would all agree that poses a
dilemma: Should I yell and keep my child safe? Or talk to him slowly
and reasonably and hope he listens? Plenty of parents would choose
yelling and obedience.

Not surprisingly, children who grow up in poverty, who sometimes live in very stressful and chaotic environments, are more
likely to have challenges with self-regulation.[19] While the nurturing
interaction of parents and caregivers can buffer children from the
adversity that surrounds them, oftentimes, despite parents' best at-
tempts to keep it at bay, tremendous stress can seep into the developing brain of a child. Constant, chronic stress elevates cortisol
levels in babies and young children, which is a problem because in
addition to triggering a host of physical ailments, cortisol changes
the brain on a cellular level—possibly altering the growth and development of the prefrontal cortex and, therefore, children's behavioral responses.[20] In other words, while children's cognitive and
executive function skills aren't *inborn,* too often children are *born
into* a world of adversity, which will have a negative effect on those
skills.

We must view parenting through the lens of the world in which
people live, the experiences they have, and the competing forces
they face. Did Don's death affect what I did as a parent? Absolutely.
It made me overprotective. To this day, each time I look at Lake
Michigan, whether it is as turbulent as on the day Don died or placid
as a mirror, I see only a threat to my children and my ability to get
them to the other side of the river, to adulthood. More starkly, parents who live in homeless shelters and high-crime neighborhoods

grapple with very real fears every single day about whether it's safe for their children just to walk down the hall or to school. Like parents of the past, they still have to think about basic survival even as they also aspire to deliver the kind of intensive parenting that Patrick Ishizuka studied, the kind of cognitively based parenting that will help their children not just survive but flourish in today's world, which usually means going to college and working toward a profession. Many low-income parents are straddling two different parenting worlds.

What Sabrina really needed was a better place to live. But again and again, her family was turned down for housing. Usually, the reason given was that Wayne had a nonviolent arrest on his record (for possession of marijuana) from a decade earlier when he was an adolescent (before his own executive function was fully developed). They were told that they would have a better chance at housing if Sabrina applied on her own as a single mother, which would mean breaking up the family or lying about their status. Think about that. Here was a married couple, raising their two children together. He had a job and hadn't been in trouble in years. But the system had no way to help them other than encouraging them to split up. "I'm not doing that," Sabrina insisted.

Some days, though, her resolve crumbled. After taking care of the basics like feeding her family, Sabrina crawled back in bed. She admitted to us that she was forgetful, struggling to remember appointments and to give her boys consistent instructions. Jason would correct her: "No, Mommy, that's not what you said." Her joints ached but she had put off going to the doctor. What was really going on was that Sabrina was depressed—quite common among mothers who have been evicted.[21] At least she recognized the problem and knew she had to get help. "I can't keep putting myself off," she told me. "If I'm not okay, how can I take care of Nakai?"

The Torrent Within

Hundreds of miles to the west, Katherine was also worried about taking care of her young son. On the surface, Sabrina and Katherine looked very different. One was Black, the other white; one lived in urban Chicago, the other in rural South Dakota; one sent her kids to Chicago public schools, the other planned to homeschool. But the undertow of mental health struggles got an equally strong grip on them both, and the possible consequences were equally frightening. Sometimes the torrent we must cross is not as obvious as homelessness; it lives within us. But whether the source of stress is external or internal, it has a formative effect on executive function and healthy brain development.

The nights after Katherine gave birth to her first son were sleepless, but not for the usual reasons. Even when her infant was settled and quiet, she'd toss and turn. Her mind was fixated on one bizarre thought: that a car driving down their road at full speed might crash through her bedroom window and kill the entire family. When Katherine rose in the morning, she found it nearly impossible to leave the house. She was terrified to let the baby out of her sight. Even when she showered, she put his car seat on the bathroom floor where she could see him. Her mother came to help and insisted on watching the baby while Katherine went to the store. But Katherine sat in the driveway and cried, unable to drive away.

Katherine became paranoid about taking her son out in public. Her husband is a pastor, and there were people in his congregation who wanted—expected, really—to meet the baby. Thinking about how she would have to pass him around a circle to let others hold him filled Katherine with dread. She made excuses and told people she was exhausted. They would smile and nod and say, "Oh, it's because you have a new baby." Katherine agreed, but she knew otherwise, that the intrusive thoughts were invading her days, too.

She wasn't just feeling anxiety. Katherine also felt immense shame. *Why am I struggling?* she thought. *I'm really good with other people's kids. I shouldn't be feeling this way. What is wrong with me?*

Katherine had a form of postpartum depression, but she didn't realize it. In the first days after delivery, many mothers experience a bout of "baby blues"—mood swings, crying spells, anxiety, and difficulty sleeping are all common and can last up to two weeks. They are caused by the extreme swings in hormones triggered by childbirth and by the exhaustion, powerful emotions, and dramatic changes in daily life that having a baby brings. Postpartum depression may look like baby blues but is usually more intense and can last for months or even a year or more. It can include a wide range of symptoms, from crying and fatigue to the severe anxiety and panic attacks Katherine experienced.[22]

At the time that Katherine gave birth, she and her husband, Daniel, were living near Sioux Falls, South Dakota. But they were new to the community. They had met on a dating site several years earlier, when Daniel was in seminary school and Katherine was living on the Pine Ridge Indian Reservation, teaching fourth and fifth grade as part of the Teach for America program, and managing an organic farm that distributed free fruit and vegetables to local families.

When she found out she was pregnant, Katherine hoped to have a home birth, but there were so few midwives in the area that a hospital birth was really the only option. Her labor was twelve hours—with no epidural and no medical interventions—and she gave birth to a beautiful baby boy. Six weeks later, she had one—and only one—postpartum checkup. It was perfunctory.

"How are you doing?"

"Fine."

"How's breastfeeding going?"

"Great."

"Do you want an exam?"

"No." Katherine didn't think she needed one.

And then she went home.

Granted, the doctor did ask how she was doing, but Katherine gave the easy answer, and without digging for more information, the doctor was never going to find out the truth. "The one-time six-week checkup was not enough to catch it," Katherine said.

Like Katherine, Daniel was a first-time parent, so he didn't know what was normal any better than she did. He didn't pick up on how fragile her mental state was. They had both heard of the "baby blues," but Katherine wasn't crying all the time, which is what she imagined baby blues would look like. Nor was she unable to care for her son, which is a common and dangerous consequence of postpartum depression if mothers neglect their babies or feel an urge to harm the babies or themselves.[23] Instead, Katherine had those intrusive thoughts and that severe anxiety. She was also short-tempered with everyone. Even though all of that was a far cry from her usual demeanor, she didn't see it for what it was, and neither did Daniel.

One of the great problems with mental illness is that sometimes it works like an invisible riptide—remaining hidden beneath the surface. And too many people try to keep it hidden there because of the unfair stigma associated with it, the fear that others will view it as a personal failing. Though one in five American adults have experienced a mental health issue—and more during the pandemic—and one in twenty-five live with a serious mental illness such as major depression, bipolar disorder, or schizophrenia, more than half of adults with mental illness do not receive mental healthcare treatment.[24] Aside from reluctance to confront the issue, or, as in Katherine's case, not understanding that what they are going through is something that *can* be treated, cost is a major reason for failing to get help. The Affordable Care Act requires insurers to cover behavioral and mental healthcare, but many people still don't have the resources to pay their share. There also aren't enough mental health professionals to go around. Over one-third of Americans live in areas deemed lacking in adequate numbers of mental health professionals.[25] Rural areas like Katherine's often have no mental health professionals at all and urban areas have long waiting lists. There are also racial barriers to getting treatment. Black and Hispanic/Latino people have a harder time both at finding services and at finding clinicians who look like them and can understand their cultural experiences.

In Katherine's case, it took her six months from the time she gave birth to acknowledge that what she was going through wasn't normal. And once she did, she didn't know what to do about it. "I never reached out for help when I should have because I didn't know where to turn," Katherine said. Just as she had finally diagnosed it herself, she treated it herself. She changed her diet and exercise, which seemed to help. One year in, she was experiencing only a few traces of anxiety and those were manageable. (Although this worked for Katherine, it is recommended that anyone who suspects they or a loved one has postpartum depression should seek professional help.)

Postpartum depression is just one example of the torrent within. But whatever the issue, a parent's untreated mental illness can have real ramifications for children. Living with a parent who is mentally ill qualifies as an adverse experience for a child *and* puts that child at greater risk of mental health issues later in life. Strong executive function skills can help a child cope, but they are also harder to develop when living with the uncertainty of a mother who struggles with depression (or a father who drinks too much or faces other mental health challenges).

How much different might that year have been if Katherine had understood what was happening to her sooner? "If people were informed about how postpartum depression could look, maybe we could catch it earlier," she told me. And if she had known more, it would have been easier to reach out for help. "Postpartum support would fill a huge void for a lot of women." In her ideal world, Katherine envisions creating intentional parenting communities, spaces that could bring together foster parents, young parents, older parents, and everyone in between.

Life Jackets

I love Katherine's vision. It would be a human safety net of sorts. But we also need stronger societal safety nets. The profound negative impact of these external and internal stresses must push us to see the societal issues that trigger them in a different frame. So often we fail to see the connection between issues like homelessness and healthy brain development, or between parents' struggles with psychological issues and their children's brain development, when in fact they are closely related. Laying the foundations for strong language and strong executive function requires calm, stable environments for children to grow up in, and calm, stable, mentally healthy parents, too.

Here, again, is the disconnect between what we know and what we do. We know postpartum depression and other mental illnesses are dangerous for mothers and fathers and their children, yet in many parts of the country—especially but not exclusively rural areas—we aren't set up to notice mental health problems, much less do anything to alleviate them. We know that a homeless shelter is a terrible place to raise a young child, yet it can take years to find a real home for a family. Is it any wonder that when children experience such setbacks in their early years, we see such stark inequities later? Sociologist Jennifer Glass, the same researcher who studies parental happiness, has found that countries with less robust family support policies are more likely to show disparities in the health of children—and those early gaps compound into larger adult inequalities.[26] To state the obvious, the countries that have more robust family policies have narrowed the health gaps in childhood, which translates into narrower gaps in adulthood.

We *know* what protects children: a positive, supportive relationship with at least one adult; exposure to rich language inputs; and a safe, stable environment. Children with a strong relationship to a caregiver are better able to regulate their emotional responses to stressful situations—that's executive function at work. Children need to be cushioned. We need to give them life jackets to keep them afloat. And that means giving the adults in those children's lives life jackets, too. When we do that, good things happen.

By the time Katherine had her second child, she and Daniel had moved to a small village in Michigan, closer to family. This time Katherine knew what postpartum depression looked like, but she also made sure she had the support she needed. She found a midwife who also became a friend. Throughout her pregnancy, the two met once a month for an hour. They did all the usual prenatal checks— listening to the fetal heartbeat, checking the baby's position and Katherine's weight and blood pressure. But they also talked about how Katherine was feeling. Instead of a brief check like she'd had at the hospital, Katherine felt she was in control and had time. "I really liked it. I knew that everything that I was doing was my choice. And, also, everything was explained to me instead of just done."

Her second labor was quick—just four hours–and smooth. She gave birth to another ten-pound baby boy at home, just as she wished. Afterward, her midwife cleaned everything up and put a meal for the new family of four in the crockpot. She came the next day to check on them, then again three days later, a week later, two weeks later, and a month later. Altogether, Katherine had eight post-partum appointments with her midwife, compared to one with her ob-gyn in South Dakota. "I felt really supported."

Indeed, after her second son was born, Katherine stayed in bed for three weeks, a practice known as a "lying-in period" that was once common. It is intended to give new mothers time to heal and bond with their new babies. It helped that Daniel had six weeks of paternity leave this time. "I just stayed in bed and hung out with the baby and had food brought to me," Katherine said. There was no postpartum depression the second time. "I think a lot of it was the way that I was treated before birth, during birth, and after birth."

Things improved for Sabrina as well. She has gone back to school to finish her degree in creative writing and English. And more than two years after they arrived at the homeless shelter, Sabrina and her family finally left. After being turned down several times for possible housing, they succeeded in finding a house through a nonprofit affiliated with their older son's middle school. Initially they were 104th on a list of families, only 100 of whom could be helped. But not everyone qualified and one day, while the family was at the library, Sabrina's cell phone rang. They had moved up on the list and were in! It still took months to find the right place, but now Sabrina, Wayne, Jason, and Nakai live in an adorable house in a neighborhood with far less violence. The new house is small, but Sabrina has put her housekeeping skills to work and made it a home. The boys can walk to school. With space to explore, and a sense of safety, Nakai is a different child.

"In the shelter, he was very guarded," Sabrina said. "He wasn't so open to taking risks. He would never talk with anybody but us. He would never engage. Being around a lot of people made him nervous." But now, Nakai hangs out the window of their new house, waiting for new people passing by. "He speaks to everyone that he comes across. He has blossomed."

Honestly, training kids to resist marshmallows is not the biggest difference we can make. We can have a more meaningful impact on children by ensuring that their families have safe places to live and that everyone's physical and psychological needs are being met. Only then can all children have a fair chance at earning two marshmallows.

PART THREE

THE WAY FORWARD

EIGHT

LIFTING OUR VOICES

"The most common way people give up their power is by thinking they don't have any."

— ALICE WALKER[1]

The events of Saturday, April 6, 1968, loom large in my family's lore. At the time, my parents were living in Baltimore, where my father was a pediatric resident at Johns Hopkins and my mother was a social worker. My mother worked full-time directing a community outreach program that was part of President Lyndon Johnson's War on Poverty. On that fateful Saturday, the country was on edge and Baltimore was, too. Two days earlier, Martin Luther King Jr. had been assassinated in Memphis. The mayor called an emergency meeting with community leaders and organizers such as my mother. The long-simmering anger and frustration of Black residents over racism and economic oppression was boiling over. The first plate-glass window—in a hat shop—was smashed around 5:30 P.M.[2]

While she was still downtown, my mother was summoned to the phone. It was my father, who was across town in the Johns Hopkins Medical Center complex. Neither of them can remember how he tracked her down, in that time before cell phones. But since he didn't usually call Mom while she was working, she knew when she heard his voice that it must be important.

"You need to come home now," he said.

The situation in the city was rapidly deteriorating. He was watching it on the news. My mother's main concern was the safety of the people she worked with and for, but she understood that her own safety was also at risk. She knew my father believed deeply in her and her work. Asking her to come home wasn't something he would do lightly.

She was also nine months pregnant with me.

Mom got in her car to drive home. Pulling into the street, she looked in the rearview mirror. The world behind her was enveloped in flames. Baltimore was burning. As she crossed the city to get home, the fires continued to flare.

Mom was frightened, but she was also frustrated and heartbroken. She recognized a piercing truth: While she was able to drive to safety, the people who lived in the neighborhood she served, and her colleagues who lived there, too, could not. Their neighborhood was going up in smoke.

She got home, had an anxious dinner with my father, went to bed . . . and went into labor. I was born the next day, on April 7, 1968.

The fires and riots lasted several more days, an expression of rage, grief, and frustration that shook the city.

Two weeks later, my mother went back to work, determined to help in the aftermath. She took me with her, something that would be unusual even today and was downright amazing in 1968. She cleared out a tiny room near her office and made it a nursery. Then she hired a young man to sit outside the room so that when I cried, he could alert her. Mom would come feed me and rock me, and then once I went back to sleep, she'd get right back to work.

*   *   *

Another mother in nearly the same time and place didn't have the option of bringing her children to work. Tragedy struck because of it. In 1965, Freddy Joyner's mother (her own name is lost to history) was living in Washington, DC, with her children. She was poor, Black, and single and couldn't afford childcare or a babysitter. Instead, Freddy, a first grader at Harrison Elementary School, grabbed the free lunch he got at school and snuck home to check on his two younger siblings and bring them something to eat. No one ever noticed when Freddy left school. The lunchroom teacher assumed he was in his classroom; the classroom teacher assumed he was in the lunchroom. It was up to Freddy to make sure his siblings were fed and safe and he faithfully did his job—right up until the day when he was crushed beneath the rear wheels of a truck at Fourteenth and U Streets while racing home. Freddy was six years old.[3]

I never crossed paths with Freddy. He died before I was born. But looking back in my own rearview mirror, I see the threads connecting me and my mother and Freddy and his—the needs of children, the needs of mothers, the tension between work and family, even the backdrop of the civil rights movement and the fight for gender equality. I also see what separated us—to put it starkly, the circumstances that determine who can drive away and who will get run over. A broken system affects everyone, but especially those living in poverty.

I also see the four of us woven into the wider context of a momentous time. "We are caught in an inescapable network of mutuality, tied in a single garment of destiny," King wrote in 1963 in his "Letter from Birmingham Jail." "Whatever affects one directly, affects all directly."[4] People felt that keenly in the 1960s. The year I was born, 1968, was dubbed "the year that rocked the world."[5] It brought world-changing disruption: convulsive antiwar protests, dramatic anti-authoritarian uprisings in Latin America and Eastern Europe, and the horrifying assassinations of King and later Robert Kennedy. But it also brought continued progress in women's and civil rights. There was a national awakening of sorts. In King's honor, President Johnson urged speedy congressional approval of the Civil Rights Act of 1968, the third and final piece of significant civil rights legislation passed that decade. It was enacted the day after King's funeral.[6] In 1968, the sparks of the movement for gender equality also ignited, lit by protests at the Miss America pageant that September. Within a few years, Congress passed the Equal Rights Amendment, with support from both Democrats and Republicans, and it looked like it would soon be ratified by the necessary number of states.

We thought that true change was coming for women and for people of color.

We thought it was coming for children, too.

The needs of children and mothers and families were an essential yet unappreciated part of the women's movement and of the civil rights movement, closely related to the prevailing concerns about work and justice and opportunity. For one glimmering moment, those needs rose to national attention. It was the death of Freddy Joyner that brought them there.

"How could this happen in America?"[7]

That was the question posed by Walter Mondale as he pondered the tragedy of Freddy's death. How could childcare have been so far out of this mother's reach? How could no one have noticed that Freddy wasn't in school at lunchtime? How could a family be forced to rely on a six-year-old as its safety net?

Freddy could have been forgotten, just another unremarked tragedy of poverty. Instead, his story became a catalyst for legislation that could have changed the course of history. The pivotal words are *could have*.

Young Senator Mondale arrived in Washington from Minnesota

the year before Freddy died. He quickly grew tired of merely talking about doing right by families like Freddy's. He wanted to act. By the end of the 1960s, he thought the time had come. Because even after a decade that had seen such change—women demanding equal rights and entering the workforce in ever larger numbers, landmark civil rights legislation prohibiting discrimination and protecting voting rights—there was little that addressed the needs of children and families head-on. No one was out there protesting on behalf of kids the way they were fighting for civil rights and women's liberation, but Mondale felt an insistent demand for action nonetheless.

He responded to the needs of children and families in much bolder fashion than anyone before him. Joining forces with Representative John Brademas of Indiana, a champion of childcare, Mondale began pushing for an ambitious bill that would deliver meaningful change and would prevent another tragedy like Freddy's. And I do mean ambitious. The bill that Mondale and Brademas ultimately brought to Congress, the Comprehensive Child Development Act of 1971 (CCDA), stated in its preamble that it aimed "to provide every child with a fair and full opportunity to reach his full potential."[8] It's a straightforward but profound goal. To achieve it, the bill called for the creation and funding of an infrastructure of support where there was none. It envisioned comprehensive and, in time, universally accessible early childhood development programs throughout the country, created in partnership with parents and communities. It would serve any families who wanted to take part, regardless of whether the parents worked outside the home. The goal was to provide not just childcare but an educational foundation based on the latest recommendations of developmental psychologists, educators, and other experts. While the bill itself may have been inspired by a boy whose family was living in poverty, it offered help to *every* socioeconomic group on a sliding scale, acknowledging that high-quality childcare and parental support should be available to all.

One of the most impressive parts of this story is the jaw-dropping amount of bipartisan support that the CCDA garnered. It also attracted unusually broad support across society, including labor, religious, women's liberation, and public interest groups. The bill had prominent Republican co-sponsors, and one of the drafters was Edward Zigler, the head of President Richard Nixon's new Office of Child Development, one of the nation's premier experts in child development and a founder of Head Start.[9] In 1971, the CCDA bill sailed through both chambers of Congress, passing the Senate in a 63–17 vote. Was it a perfect bill? No. Did everyone support it? No. But it was an extraordinary moment of political consensus, something that seems unthinkable given the divisions in Congress today. *The Washington Post* called the bill "as important a breakthrough for the young as Medicare was for the old."[10]

Then, in a stunning about-face, Nixon vetoed his own bill.

This was a president who had promised to provide "all American children an opportunity for a healthful and stimulating development during the first five years of life." He had established the Office of Child Development that Zigler was leading. And he headed a Republican Party that advocated for better programs for preschool children in its platform.[11] But by the time the CCDA made it to his desk, Nixon had rescinded his previous support for the bill. Why?

Because of politics. Despite the bipartisan support for the bill and Nixon's initial embrace of it, a small group of his advisors, led by the conservative Patrick Buchanan, thought the bill had to be stopped.[12] They lobbied Nixon hard to change his mind, arguing that a federally supported childcare system looked suspiciously like what existed in the Soviet Union. The president was torn, but in the end, he was convinced that signing the bill, on top of his visits to the Soviet Union and China, risked alienating anti-Communist, "family values" voters. (Remember, when the Soviets launched the Sputnik satellite in 1957, we took the opposite approach and ratcheted up our science and math education. We did not abandon the effort altogether.)

The veto on its own was devastating. But Nixon compounded the damage with the speech, written by Buchanan, in which he announced his decision. He used such strong language that he "drove a stake through the heart" of the ideas in the CCDA, shattering them into a million pieces, in the words of one commentator.[13] Those pieces mutated into the chaotic patchwork of dysfunction that marks childcare in the United States today. Despite the fact that so many wanted the CCDA and would have benefited from it, Nixon successfully deployed the powerful trope that governmental support of families is a direct assault on American individualism, the sanctity of family, and the rights of parents. Of course, parental choice should be cherished and protected. But, in actuality, vetoing CCDA slammed the door on what would have been a wide range of options. Signing it? Now, that would have given parents true choices. In truth, the bill included provisions not just for parents working outside the home but for mothers who wanted to stay at home with their children. (Given the "choice," Freddy's mom would undoubtedly have chosen almost anything other than having her six-year-old run home from school to care for his young siblings while she was at work.) The Nixonian version of parental choice was more symbolic than real, and it has impeded progress for the five decades that followed.

Treading Water for Fifty Years

Fifty years have passed since the brief life and dramatic death of the CCDA. Not nearly enough has changed for families in that time. Nixon's message rolled like a chilly fog through the subsequent decades, making it hard to move forward with meaningful progress and obscuring the real costs of inaction. As other developed nations enacted policies like paid family leave, universal childcare, and home visiting for new parents, the United States stood out for its failure to adopt any such policies or programs and for spending less on early childhood than just about any other developed nation.

But the veto of the CCDA also exacted a more insidious toll. It undermined the powerful promise and hope of the civil rights and gender equality movements. The fate of children and parents is one of the threads holding together the "garment of destiny." When those threads unravel, the fabric of society is frayed.

If children of all races, ethnicities, and genders are to flourish and grow up to participate equally in the economy and the civic life of the nation, society must support what is required for healthy brain development from birth. That is how we grant children the promise of their innate promise. That is how we give the people who care for those children, especially mothers, a fighting chance. Healthy brain development is a fundamental condition for equality—a basic human right. Without it we will never be able to create true and lasting change.

We can get a glimpse of the kind of change that *is* possible when family and child support are made available from the famed Abecedarian preschool project, which launched one year after the CCDA was scrapped. Abecedarian was a careful research study involving 111 children born in North Carolina between 1972 and 1977.[14] All the children and their families received healthcare, social services, and nutritional supports from infancy through kindergarten. In addition, roughly half the children received five days a week of high-quality childcare year-round, centered on language ("Every game is a language game," the project's creators said).

The purpose of Abecedarian was to investigate whether quality early education could improve school readiness for babies from low-income families. And it did. In their school years, the Abecedarian babies who attended childcare, most of whom were Black, did better in reading and math than the children whose families received only health and social services. But that was not all.[15] At twenty-one, the childcare participants were more than twice as likely to have enrolled in a four-year college and less likely to have been teen parents.[16] As adults, they had lower rates of obesity and high blood pressure and were less likely to engage in criminal behavior. They also earned more money.[17] As for the mothers whose children were part of the preschool program, they didn't just benefit indirectly from the boosts to their children's prospects. Those who had been teenagers when they had their children were more likely to have finished high school than mothers with children in the comparison group. Like their children, they were also likely to be more prosperous. A 2007 cost-benefit analysis conservatively estimated they earned about $3,000 more per year than they would have if their children had not participated.[18]

A half century after the veto of the CCDA and after Abecedarian began, it is hard not to see the parallels between this moment in time and the dramatic events of the year I was born. Still, American citizens feel the sting of racism and the injustice of inequality. Once again, there have been riots and civil unrest. The trauma of COVID-19 has been global and sweeping, but it did not affect everyone to the same degree. The accounting of who suffered most, and why, offers further, dramatic proof of how intermingled the issues of civil rights, gender equality, and family needs really are.

For parents of young children—but especially mothers—the challenges the pandemic brought to everyday life were unsustainable: closing of schools, the normalizing of remote work, and the crumbling of an already fragile childcare system. Mothers were more likely to shoulder the pandemic's additional childcare responsibilities. They were more likely to lose their jobs or to have to quit to care for family members. Mothers were also more likely to suffer worsening mental health. "Covid took a crowbar into gender gaps and pried them open," said economist Betsey Stevenson of the University of Michigan.[19]

The pandemic reminded us that, sometimes, desperate parents have to leave their young children alone, just as Freddy Joyner's mother did. I heard about a young father and custodial worker in Oklahoma who had to take his two children—six and four years old—with him to work every day. Because COVID restrictions meant the children couldn't come inside, he left them locked in his truck in the parking lot all day, checking on them once every hour.

Of course, parents faced impossible choices even before the pandemic. At a church-run parent support group attended by an acquaintance of mine, a newly divorced mother of three children under the age of ten (I'll call her Tatiana) shared her story. Her ex-husband was supposed to have the children one weekend, so Tatiana signed up for double shifts at the hotel where she was a housekeeper to make some extra money. But only hours after picking up the kids on Friday, her husband returned them to her door. The four-year-old had gotten sick to her stomach. "I cannot deal with this," the father said. "This is your job!" With no friends or relatives available, Tatiana called her boss to say she couldn't work, but her boss said she would lose her job if she didn't show up. After a mostly sleepless night, Tatiana decided she had no choice but to leave the children alone. She delivered strict instructions not to leave the apartment and called every thirty minutes while she was gone. "Please don't judge me," she told the support group tearfully. And they didn't. Instead, several of the other parents, who had faced their own hardships and parenting challenges, immediately handed over their phone numbers.

While the pandemic was hard on all parents, it was especially hard on Black, Brown, and Native American parents, who contended with both the crisis in childcare and education and with entrenched racial inequities that heightened their risks of getting ill and dying. No matter where they lived or how old they were, people of color were disproportionately affected by COVID-19. Black and Latino people were significantly more likely to get infected than their white neighbors and they were also more likely to die, according to the Centers for Disease Control.[20] The reported rates of infection and death of Indigenous people were similar, although the real rates are thought to be much higher. Many of the reasons for these elevated risks—frontline jobs, reliance on public transportation, crowded housing, less access to healthcare, underlying health conditions—stem from our country's history of racism.

That history reared its ugly head in May 2020, during one of the worst phases of the pandemic, when the world witnessed the murder of George Floyd in Minneapolis and erupted in collective rage. Floyd's death forced a national reckoning with the myriad ways that Black people are still treated unfairly in our society, particularly by the police. In the wake of Floyd's death, there were weeks of protests. An estimated fifteen to twenty-six million Americans participated, making the Black Lives Matter movement one of the largest— probably *the* largest—in American history.[21]

The riots in the wake of Martin Luther King Jr.'s assassination should have been a once-in-a-lifetime event, but the scene my mother left behind in Baltimore in April 1968 could have been from May or June 2020.

Certainly, the dire economic situation that was one cause of the unrest of 1968 in Baltimore has not improved. Segregation, poverty, and homicide rates have stayed much the same there since the 1960s.[22] Between 2008 and 2012, in one West Baltimore neighborhood, 51.8 percent of the residents were unemployed and the median income was just $24,006 per year, a mere $723 above the poverty line. In Baltimore as a whole, 58 percent of students are low-income, a statistic the school system says under-reports the true number.[23] When asked by researchers from Johns Hopkins, Baltimore teenagers reported feeling worse about their circumstances than their peers in New Delhi, India, and Ibadan, Nigeria.[24]

Baltimore is not an anomaly. I think of the gunshots that rang out in Randy's Chicago neighborhood. Or the fact that the homeless shelter that Sabrina and her family had to turn to was hardly "shelter" at all. I imagine Hazim Hardeman's mother riding the bus to a distant neighborhood, willing to lie about her address and risk her own freedom to put her sons in a safe and stimulating school. These are not hurdles parents can easily overcome.

How could they? Parents are regularly reduced to tears over the less dramatic but still painful choices that confront them. Kimberly Montez wept when she was unable to be with her baby in the NICU. And Jade cried both when she had to leave her son to go to work and, again, when she told me about it, years later.

Is this the best we can do?

We don't always know what is possible or what to ask for. Sometimes we are limited by where we set our sights. Then, suddenly, we see over the horizon and understand there is another way. I was reminded of that fact during a conversation with a friend and colleague here at the University of Chicago, Ellen Clarke. I was telling Ellen how learning the stories of some of our TMW families and the difficulties that these parents faced in their children's earliest years had led my thinking to evolve. It turned out that Ellen had been thinking about the context in which we parent for different reasons. She had been watching her own little social experiment unfold.

In her mid-thirties, Ellen is in a phase of life when she and many of her friends are having children. Tracking the different experiences of her group of friends has been eye-opening. Several of them grew up in the same small Wisconsin town and went to the same university. But from there, their paths have diverged dramatically. Neither of the two friends who stayed in the Midwest—Kristen, a scientist, and Ashley, a nurse practitioner—got paid maternity leave. Their husbands didn't get much leave either, even when one of the children spent six weeks in the NICU, a painful repeat of Kimberly's experience. On the other hand, Ellen's friends Diane and Rebecca relocated to Norway and the Netherlands, respectively, because of job opportunities, and they had babies there. In both countries, the women and their husbands got generous parental leave they could schedule as they preferred. Furthermore, they were encouraged to take every day they were entitled to. Once they went back to work, they were able to turn to the strong Dutch and Norwegian childcare systems.

Ellen shook her head as she told me about all this. Like Gabby, she herself had struggled to find childcare that was conveniently located, that was affordable, and that had a spot open. She jokes about childcare being the budgetary equivalent of "a second mortgage." She and her husband often feel they are scrambling just to hold steady financially and emotionally. Yet Ellen knew that by American standards, she had been "lucky." She had been able to take a fourteen-week paid leave, her husband had some flexibility in his schedule, and they could afford decent childcare. "Lucky" didn't feel like quite the right word, though. "I felt like I shouldn't have to feel lucky. This should be the bare minimum of what families are able to do," Ellen said. Certainly, her friends in Europe with more institutional supports were markedly less stressed. Is it so surprising that countries with more robust family support policies have narrower health gaps and happier parents (and non-parents)?

The unrealistic expectations placed on American parents are crushing. "We were told that we should take this all up on our shoulders," Ellen told me. "You need to own up to the idea that you *shouldn't* have to. Everyone is at home, underwater, by themselves and not able to raise their hand."

I must admit that when I heard her say the word "underwater" it took my breath away. I thought immediately of Don's death in Lake Michigan. And I thought of my dream. I saw myself standing alone on the shoreline. And, this time, I saw the parents I've met standing there, too. Each was alone just as I had been, carrying their challenges on their own shoulders, readying their own boats. Each was completely unaware of the others standing next to them.

No Place for Complacency

There was a time when we successfully pushed forward a monumental, sweeping public health program for children and women in this country. It was inspired not by the death of one child but by the death of many. In the early 1900s, birth in America was a perilous endeavor. Infant mortality rates were high. The great majority of mothers received no advice or trained care during their pregnancies or labor—a problem that was especially pronounced in poor and rural areas. (As we saw in Katherine's case, rural women still have a hard time getting medical care).

But then several promising developments converged. In 1912, the United States established the Children's Bureau, the first federal government office focused solely on the well-being of children and their mothers.[25] It was also the first government agency headed by a woman, Julia Lathrop. Then, in 1916, Jeannette Rankin, a Republican from Montana (note that she was from a rural state), became the first woman elected to Congress. Finally, in 1920, American women were granted the right to vote. Rankin and Lathrop joined forces to provide support to mothers and their infants. In 1921, a version of a bill that was written by Lathrop and originally introduced by Rankin was signed into law.

Known as the Sheppard-Towner Act (because Rankin had left Congress by the time it passed, it was named for its sponsors,

Senator Morris Sheppard of Texas and Representative Horace Towner of Iowa), the bill established thousands of prenatal clinics, provided for millions of home visits by traveling nurses, and greatly improved the grassroots educational information available to mothers about maternal and infant health. It demonstrably saved lives: Infant mortality rates dropped dramatically in the states that participated most fully in the program—for example, deaths associated with gastrointestinal diseases, a major focus of the educational programming, fell by 47 percent over the life span of the bill.[26]

The key to getting such a bill passed was women's newfound political power. Politicians were afraid of women's power as a voting bloc and they were afraid of voting against something like Sheppard-Towner. "For years, suffragists had promised to clean house when they got the vote," wrote one historian.[27] Politicians took them at their word. But five years later, when the initial funding had expired, it had become clear that women didn't vote as a bloc. Male politicians no longer feared opposing an extension of the law. And opponents like the American Medical Association, whose members didn't want to see the use of nonmedical providers for medical services and viewed the act as moving the country closer to socialized medicine, were better organized the second time around.[28] (Notably, the pediatric group within AMA supported the bill and when the AMA lobbied against it, the pediatricians split off and created the American Academy of Pediatrics in response.) In the end, there was a compromise: Congress extended funding but only for two more years.

It's those kinds of shifting political winds that should give us pause today. Political coalitions are fragile. Public attention is fleeting. But the pain and struggle of families is real and enduring. Infant mortality rates are *still* high among Black mothers.[29] And families suffer in other ways. It's hard not to think of those kids in that pickup truck in Oklahoma. No child should have to spend the day in a parking lot.

Learning from Our Elders

And no one should have to live in a chicken coop.

Yet an elderly woman once did. And the discovery of her there, in such desperate circumstances, ultimately changed our nation for the better. It's a story that has much to teach us about how parents might bring about the real, essential change that will allow our children to flourish.

In the mid-1940s, a woman named Ethel Percy Andrus was a member of a California state committee concerned with the welfare of retired teachers.[30] That was a category Dr. Andrus had recently joined herself, after four decades of being a celebrated teacher as well as the first female urban high school principal in California and a proud University of Chicago graduate. When news of Andrus's work on teacher welfare was published, she got a call from someone asking her to check on a former teacher. Andrus drove to the address she'd been given, but no one answered at the house. Just as she was about to leave, a neighbor suggested perhaps she was looking for the woman "in back." Behind the house, Andrus knocked on the door of a windowless shed, a former chicken coop. It was answered by a woman in a threadbare coat who slipped outside to talk. To her shock, Andrus realized that this was a woman whose name she knew, a "Spanish teacher of some distinction." The woman had planned for retirement by investing in land, but the Depression and then a flood on the property ended those hopes. She was making do in the coop—all she could afford on her meager teacher's pension.

Andrus was so upset by what she saw that she spent the rest of her life working toward one mission: ensuring that no retired person be reduced to living in such conditions.

She had her work cut out for her. In the mid-twentieth century, Americans over sixty-five years of age were the poorest, most underserved segment of the U.S. population. In addition to having limited retirement savings, older Americans faced crippling healthcare and housing costs. According to Michael Harrington in his 1962 landmark book on poverty, *The Other America*, "fifty percent of the elderly exist below minimum standards of decency."[31]

Andrus's first step was to create the National Retired Teachers Association (NRTA). The organization worked toward pension reform, tax benefits, housing improvements, and health insurance. Health insurance was virtually unavailable to older Americans at the time. Andrus approached forty-two insurance companies and forty-two said no. "They thought I was a crank," she said. But finally, she found one willing to take on the risk of insuring older people. The NRTA, however, served only retired teachers, and Andrus had much broader goals. In 1958, she joined forces with others to expand the scope of the organization and found the American Association of Retired Persons (AARP), an organization dedicated to the support of the entire population of older Americans.

Their success is legendary. Thanks to AARP's efforts over the last fifty years—and to the voting clout of the tens of millions of people in that age group—there is no constituency better served by society and government than the elderly. The poverty rate among Americans aged sixty-five and older has declined by almost 70 percent.[32] During the sixties, the Social Security Act, which FDR had signed into law in 1935, underwent numerous changes that expanded the benefits available to older adults. The passage of the legislation that established Medicare and Medicaid in 1965, and of the Age Discrimination in Employment Act in 1967, further bolstered the economic well-being and security of that age group. Today, AARP continues to make advances in healthcare, prescription drug support, long-term care, and more. It is an organization that manages to unite constituents across socioeconomic, political, racial, and ethnic divides by focusing on rights that benefit everyone. The organization is one of the few consumer-advocacy groups that can go toe-to-toe with big corporate interest groups, wielding the immense political power conferred by their big budget and their thirty-eight-million-strong membership. No substantial policy conversation about citizens over the age of fifty can take place without AARP leadership at the table. What began as an organization designed to protect and support our children's teachers evolved into an advocacy group that has exponentially improved the lives of all elderly Americans. When a group of people speak with one voice, it's amazing what they can accomplish.

I see another lesson in Andrus's work. From the start, her message centered not just on the needs of older people but on the contributions they had made—and could still make—to society. The former Spanish teacher was in need of help, but she had also spent the bulk of her life helping others, educating hundreds, perhaps thousands, of children. And she could still be useful. Andrus didn't just push for benefits for the elderly, she mobilized millions of them as volunteers. In doing so, she changed society's perspective on old age. The motto she adopted was this: "To Serve, Not to Be Served."[33]

Cradle of Our Future

Parents do not just contribute to society, they create it. They are raising the next generation, the next wave of students and teachers, employees and employers, voters and parents. As guardians of their children, parents are nothing less than guardians of our future well-being.

Tragically, the plight of some parents and young children today is not so dissimilar to that of the elderly prior to the advent of AARP. In fact, as I have mentioned, the poorest segment of the United States population today is no longer the elderly but children under five years of age. Let me repeat that: Children, from newborn infants to five-year-olds, are the poorest of our citizens. Of course, practically speaking, this means that these young children are living with impoverished parents and caregivers. Like the elderly once were, parents are in many ways invisible, marginalized, and struggling. Just as we dramatically improved the quality of life of the elderly, we can do something similar for our youngest citizens—by first helping their parents.

Parents have already begun lifting their voices on their own behalf. Across the United States and around the world, there are groups who have been working tirelessly on the myriad issues that confront children and parents.

They are united by what they share: the experience of being parents. Yes, each family shoulders unique challenges, experiences, and strengths; but nearly all mothers and fathers know sleepless nights and overpowering love. We have all felt the desire to be home with a newborn, and the stress of finding a babysitter we trust. We worry over hitting milestones and we wonder over new accomplishments. Above all, we share the desire to give every opportunity to our children, the desire to get them across the raging torrent.

And parents have strength in numbers. As of 2020, there were sixty-three million parents with children under age eighteen living at home, according to the Census Bureau. Those sixty-three million have the potential to be the largest special interest group in the country.[34]

Their special interest is children. No one cares about our children the way parents do. Since our children don't vote and they can't contribute to campaigns, they need us to speak up for them. It is our responsibility to fight for them and for the policies that will contribute to their healthy development.

We have done it before, galvanizing around specific issues like polio and drunk driving. And we're doing it today. I see groups of parents everywhere trying to affect change. For several years before the pandemic, families from all fifty states and the District of

Columbia descended on Washington, DC, for a one-day event called Strolling Thunder.[35] They rallied in front of the Capitol and brought their strollers, en masse, into the building to visit their elected representatives. One dad even changed his baby on a Congressperson's conference room table. That made quite an impression! Families showed up to fight for the issues that were affecting them personally. Jessica and Sam Hibben, of New Mexico, were there in 2017 to talk about their son, Rafe, who was born with severe vision problems and developmental difficulties. They struggled to get him enrolled in Medicaid, even though he qualified because of his special needs, and they took their story to their senator, Mo Udall, who then spoke about Rafe on the Senate floor while working to prevent changes to the Affordable Care Act. Elizabeth Kehret, whose child has cancer, spoke to Iowa's senators and representatives in 2019 about the need for paid family medical leave. "Creating the space for a parent to come in and talk about what it's like to raise their baby, with their baby on their lap, is powerful," says Elizabeth DiLauro of the advocacy group ZERO TO THREE, which organizes Strolling Thunder, which is now an annual event. "These meetings are about parents, they're about families, they're about their stories."

Parents are working locally, too. In Multnomah County, Oregon, which includes the city of Portland, a grassroots campaign led by parents was instrumental in making universal preschool a reality. In 2016, a Parent Accountability Council (PAC) began meeting and organizing to formalize its vision of preschool for all. Four years later, in November 2020, it helped get that vision passed into local law.[36]

Just as there was an awakening in 1968, there seems to be another awakening now. Parents have long felt that raising children in the early years is an impossible high-wire act. And they thought their struggles were solitary. Now they are looking around and noticing they are not alone. They are nodding at one another in recognition. When parents repeat the same stories of struggle across the country, it becomes obvious that the problem is not personal; it is systemic. Systemic problems require systemic solutions. I've seen a growing conviction that society must play a critical role in supporting parents—and that support is neither an assault on parental choice nor an invasion of the family. The opposite is true. Society's support offers the true freedom of choice that every parent deserves. It is thrilling to watch families coming together, finding their collective identity, and elevating their expectations of what society can and should do.

Parents also see their rightful place as the primary architects of children's brain development. In order to build the best early learning environment for children, parents must first feel supported in their own lives. This is neither selfish nor trivial. One's health, social integration, career support, financial security, and access to community resources all have a profound impact on personal well-being and, subsequently, the ability to nurture children's development, to nurture the next generation. Private networks of friends, family, and other individuals can help get us through the day, but they are not enough to overcome all the challenges before us or the gaps left unfilled by our institutions. They are not enough to build a better tomorrow for our children or our children's children.

A one-size-fits-all solution that will address the needs of all families does not exist, but if society values the role of parents, the path forward requires that our institutions step up and better support parents and children. When society fails to support parents, our children get shortchanged. To endow our children's futures and our own future as a nation, we must invest in children today, from birth.

Ethel Percy Andrus once said, "If we are not content with things as they are, we must concern ourselves with things as they might become."[37] We must now concern ourselves with who our children can become and with the society we can become. We must fulfill our promise and our promises. We must create a parent nation.

NINE

JUST WHAT THE DOCTOR ORDERED

"An ounce of prevention is worth a pound of cure."

—BENJAMIN FRANKLIN[1]

I n the 1980s, Peter Fleming was a pediatrician in the emergency room at the Children's Hospital in Bristol, England. It could be heartbreaking work. Almost every week, a baby was brought in dead or dying. Mother after mother described finding their babies in the same way. "He was so still," they would say to Fleming between tears. "When I turned him over, I knew he was dead."[2] Officially, the cause was sudden infant death syndrome (SIDS). But that was just a label. What was the actual root cause of SIDS? No one knew, and therefore no one knew how to save these babies. "Lots of people still thought that the parents must be at fault in some way if their baby had died," Fleming said later. "They were made to feel very guilty."[3]

After seeing so many of these parents, Fleming, who was not only a pediatrician but a researcher, was determined to do something. So were the families. Led by a woman who had lost her grandson, they had set up a foundation to try to explain the mystery of SIDS.

Fleming joined forces with their organization, now called the Lullaby Trust, and got to work looking for an explanation. The answer turned out to be right there in what the mothers were telling him—though it took him some time to see that.

Fleming began by collecting information from every family systematically.[4] He even visited the places where the infants had died—a revolutionary idea at the time. Sitting on one living room couch after another, he listened as grief-stricken parents told him their stories. For a long time, an empathetic ear was all he had to offer. Then, in 1987, Fleming launched a formal study, comparing the circumstances of each baby who had died with two others the same age and in the same neighborhood who had not. Fleming and his colleagues suspected overheating, heavy wrapping, breathing problems, or infection could all be involved, and they included questions related to each. Around the same time, an Australian pediatrician was advancing the idea that putting babies down to sleep on their stomachs was the problem, even though the standard advice to parents in most developed nations, including the UK and the U.S., was to do just that. Tummy sleeping was what Dr. Spock advised and it was what pediatricians advised. Fleming admitted he included a question about sleeping position in his study only so that he could rule it out.

Two years later, Fleming's results were in. To his astonishment, the single most important factor contributing to SIDS was the baby's sleeping position: Babies who slept on their fronts were almost *ten* times as likely to die as those who slept on their backs.

The medical world was highly skeptical. Fleming understood. "I couldn't believe anything could be that simple," he said. His colleagues demanded more evidence, and he wanted more himself. So, Fleming planned a much larger, more rigorous study specifically comparing both sleeping positions. But when they tried to enroll babies in the study, Fleming's team found that word of his earlier results had already spread across the Bristol area and there weren't enough babies who were sleeping on their fronts anymore to make the research feasible. Then, lo and behold, as the local parents engaged in this natural experiment, the local rate of SIDS fell dramatically. Within three years of Fleming's initial study, it had fallen by more than half. Yet nationally and internationally, doctors were unconvinced or unaware of Fleming's work.

Then fate stepped in to spread the word about his findings. One day in 1991, Anne Diamond found her four-month-old son, Sebastian, dead in his crib. On that sunny July morning, Diamond had put her baby to sleep on his tummy, just as she had been told to and as she had done with her two older sons, James and Oliver. Like some two thousand other British babies that year, Sebastian had died of SIDS.[5]

As it happened, Diamond was a well-known reporter and presenter for the BBC, the UK's premier news organization. Like any good investigative journalist, she began digging through everything she could find about SIDS and learned about Fleming's work as well as a study from New Zealand that corroborated his findings. And then, like many grieving mothers before her, she started ringing alarm bells. But her alarm bells, unlike theirs, were heard. Because she was a household name, she was able to use her status to launch a crusade to get the British health authorities and pediatricians to officially change the recommendation that babies should sleep on their fronts.

Six months after Sebastian died, the first "Back to Sleep" campaign, a public health effort embracing Fleming's work, was launched and spread through pediatricians and other healthcare providers. The effect was dramatic. In Britain, the number of babies dying from SIDS ultimately plummeted by close to 90 percent. Similar campaigns in the United States and elsewhere had equally spectacular results.[6] SIDS still occurs and we still don't know why entirely—the main remaining risk factor is now thought to be maternal smoking—but tens of thousands of babies' lives have been saved by the combination of Fleming's research and the public education campaigns that followed around the world.

Nearly two decades later, Fleming reflected on the lessons for doctors: "the importance of being open-minded . . . the importance of working closely with the whole healthcare team and most important of all, listening to what patients tell us."[7]

I think the story of SIDS has even more to teach us. It shows the power of educating parents, especially when conventional wisdom needs to be updated, and that doing so can dramatically change outcomes for children. The thing is that all parents start as novices, and babies don't come with instruction guides. Of all the parents I have known, including myself, I can't remember one who said from day one: "I've got this." Instead, our darling newborn sits in a baby carrier (the one we just worked out how to use) in the middle of the living room or the kitchen or the bedroom. We stare at this precious new life and think: Now what?? How am I going to do this??

That's just how it is when you're a new parent. There are bound to be gaps in your knowledge and experience no matter how many books you've read and parenting influencers you've followed, or even, like Mariah, how many other people's babies you have cared for. Our society trumpets the idea that parents are their children's first, best teachers. Yet we do not have the mindset nor the infrastructure to supply the parents themselves with the knowledge, skills, and support they need to excel at that job. There is a yawning gap between what we say parents should do and what we do to help them do it.

To build a parent nation is to reimagine a society oriented around robust support for the first years of life, to look to the North Star of healthy brain development as our organizing principle. Just as laying the foundations for healthy development means connecting parts of the brain, laying the foundations for a parent nation means connecting parts of society that rarely intersect.

Healthcare is one of those parts, one with enormous reach and potential. The irony is that more than ten years ago, I stepped outside of medicine in order to figure out what needed to be done to give all my patients the opportunity to fulfill the promise of their promise. Now I find myself back in the very place where I started, because I realize that healthcare is a critical piece of the solution I was seeking.

Our systems for nurturing health and education in the early years have been siloed. Over here in one field, there is the fragmented and extremely limited world of early childhood education. Until children reach kindergarten, there is no consistent education and care system that touches the lives of most families.[8] And then off in another field, there is healthcare—obstetricians during pregnancy and pediatricians after a child is born. Only rarely do those visits include any discussion of brain development. Parent education is a necessary precursor to strengthening children's early education, but our system provides no way of reaching and educating all parents on this fundamental issue. Furthermore, these doctors' visits are usually disconnected from other supports and services.[9]

Starting Upstream

As I think of the families I've met along my own journey—the families described in this book, my friends, and relatives—it is obvious to me how much every parent could be helped by a more integrated system and by clinic appointments that are both broader and deeper. Such a change would offer a real opportunity to close the gaps between what we know very young children need to get the best start in life and what we do to foster their growth. In the effort to build a system that supports parents as children's first and most important teachers, healthcare is an obvious place to start because nearly all families are already connected to it. Roughly 90 percent of young children see a pediatrician at some point and the great majority of mothers use the healthcare system when they are pregnant and delivering their babies.[10] Just as school systems provide a way to reach all children later, healthcare is the system that could provide nearly universal access to parents in the early years. I came to understand the importance of this in my own work. TMW began by working with one parent at a time in home visits. But later we realized that we could reach far more people by embedding our programs in the places parents already were—at the ob-gyn, on the maternity ward, or at the pediatrician.

But to make healthcare a truly effective and integral part of a parent nation, we must start upstream, a concept that is gaining in popularity. To paraphrase an idea that has been expressed by others, if we saw children in a canoe heading for a waterfall, what would we do?[11] Would we wait below the cascading water and tend to them after they plunge to the bottom? No. We would wade in or throw a rope or do *something* to try to stop them from going over the edge in the first place. Even better, we would walk back along the riverbank to the spot where people are launching their boats and help them prepare for the journey ahead.

What would this look like in practice? It would mean including education about brain development at every opportunity from the first prenatal appointment on. Insofar as most parents know anything at all about brain science, they tend to learn it much later.[12] Randy and Mariah, for instance, learned about brain development because they happened to sign up for TMW, when their children were toddlers. But imagine if Randy hadn't spotted a flyer for TMW on a bus ride. Most parents can't count on that kind of chance encounter. However, most parents *do* take their children to the pediatrician, and pediatricians, nurse practitioners, and other providers are well positioned to give parents confidence in their role as brain architects.

Starting upstream would also mean looking beyond symptoms to the root causes of good health and ill health—including the social, economic, and environmental factors that influence child development. For Sabrina, this kind of care would have meant that her pediatrician or ob-gyn's office picked up on her housing instability earlier—perhaps early enough to make a difference—by asking a few relevant questions as part of a routine screening (e.g., Was there a time in the last twelve months when you were not able to pay the rent on time? How many places have you lived in the last twelve months? Has the baby lived in a shelter or been homeless since birth?). Studies have shown that families who answer yes to any of these three questions are at risk for poorer health and developmental outcomes, especially for children under two.[13] If doctors discover housing instability, we should make it easy for them to know where and how to refer families to the appropriate supports. My dear friend Stacy Lindau, a physician and researcher at the University of Chicago, did that by founding NowPow, a referral service connecting people to community resources for care.

We must also take a universal, tiered approach to providing patients with both parent education and broader services. We should start by casting the net as wide as possible and then connecting individual families to further information and social services as needed. Our SPEAK studies show that there is huge variation in how much individual parents know about healthy child development.[14] And, of course, there is an equally wide range in the support that parents need more generally. Sometimes there are immediate issues like Katherine's postpartum depression and sometimes there are more sustained challenges like a child with special needs.

Healthcare that is both universal and personalized avoids the dangers of making assumptions. Everyone who has worked in this way, myself included, has learned that we cannot know, based on appearances, when someone needs help . . . or when they don't. Some single teenage mothers turn out to have sturdy family support and boyfriends who are devoted fathers. On the other hand, a colleague told me the story of a pediatrician who instructed a resident to do a screening of a mother with three young girls, all of whom were dressed in matching pink smocked dresses complete with monograms. To the young doctor, the family screamed "money." When he emerged from the clinic room, the pediatrician reviewed the answers he'd elicited during the screening and she asked him what the woman had said about food insecurity. He admitted he hadn't asked. So she sent him back to do a more complete screening. As soon as he asked about food, the mother broke down into tears. Just that morning, she had signed up for federal nutrition benefits because her husband had abandoned the family and cleaned out their bank accounts.

At the heart of this reimagining of healthcare is a team-based approach. No one expects individual doctors to do all of this on their own, but we can bolster the team of professionals who interact with families all along the way. I envision a healthcare system that works like a larger version of my A-team in OR4. Each participant brings expertise and has a clearly defined role, but they are also capable of backing one another up and catching what might be missed.

At its most holistic, healthcare can be a hub for parents. First prenatal and then pediatric offices can be the places where a child's health and education connect. They can be places that anticipate what parents and children need, that help parents understand their role as brain architects, and that give them assistance in surmounting any barriers that are holding them back. I'm not the only person who thinks about this possibility—far from it. Across the country, there is a groundswell of interest in the idea that healthcare can become the missing link in the educational continuum and can build a bridge from the first day of life to the first day of formal schooling.

Reimagining Healthcare

To get there, we need a major shift in mindset: We need to change our beliefs about what the healthcare system should be. Healthcare in the United States has traditionally been about "sick" care. Business as usual in medicine seeks to treat disease rather than build wellness, to react to crises and illness rather than try to prevent them.[15] In addition, more and more doctors are specialists (I am one myself).[16] Although specialists are the experts you want when you need a cochlear implant or the most effective cancer treatment, we focus primarily on only one element of a patient's condition rather than their holistic well-being.

To be clear, pediatricians are specialists, too, but they are more holistic than many other doctors and they have a long history of fighting for families.[17] (Full disclosure: Not only was my late husband a pediatrician [surgeon], my father, brother, and cousin are all pediatricians, too.) The American Academy of Pediatrics (AAP) has long been an advocate for public health approaches to children's health—it was founded, after all, as a protest to the American Medical Association's opposition to the Sheppard-Towner Act supporting maternal health. The AAP recommends a string of well-child visits—seven in the first year of a baby's life and five more by the third birthday. These twelve visits occur during the three most critical years of brain development and offer vital opportunities to assess a child's growth, deliver immunizations, and, theoretically, answer parents' questions. "This is such a unique opportunity," says my friend, psychologist Rahil Briggs, the national director of HealthySteps, a program of the nonprofit ZERO TO THREE, which partners with pediatric providers to add developmental specialists who cover all aspects of families' needs. "So much change happens in early childhood."[18]

Still, of the fifteen minutes in an average well-baby checkup, only three—three!—are devoted to what is called anticipatory guidance, or information about the changes that will occur, physically, emotionally, and developmentally, from one visit to the next.[19] Three minutes for *all* those things is not a lot when you consider how much there is to talk about—motor skills, sleeping habits, best practices for feeding solid foods, using the right car seat, and on and on and on. Three minutes for *all* those things is not a lot when you consider this is making up for the lack of a baby instruction guide.

Too often, parents wind up leaving the pediatrician's office—a trusted authority on their child's health—having heard virtually nothing about the brain or overall development. When we did our SPEAK study and talked to new parents at the hospital and in the waiting rooms of pediatricians' offices, we asked them directly: Has your doctor ever talked to you about your baby's brain and the importance of early language? The answer was almost always no. Only one-quarter of the parents reported receiving information about brain growth. Even fewer heard anything about how infants learn (13 percent) or about their learning to talk (9 percent).[20] This is often because parents are focused on other things, and pediatricians are trying to serve as many families as possible and are limited by how their time and services are currently reimbursed by insurance companies.

But here's the thing: Two out of five children younger than age five experience language delays.[21] That's nearly twice the rate of childhood obesity, which is widely recognized as a public health epidemic. The medical world has finally acknowledged that language disparities are a matter of public health as well. In 2014, the AAP issued a landmark policy statement that elevated literacy promotion, and, therefore, language interactions and the brain development they stimulate, as a key component of pediatric care, and recommended public funding to support this medical practice. Its goal is to integrate information on healthy brain development within pediatric visits.[22] But so far, that is not happening in most places.

Our current approach doesn't limit only what parents get out of each visit, it limits, on a much wider scale, what Americans get from their medical system. In what is known as the healthcare paradox, the United States spends more on healthcare and gets less than nearly every other developed country.[23] It's a situation one commentator wryly likens to "the football player who may not be big but is slow."[24] The amount of money the United States spends on health and medical services, on surgeries and medications, on doctors and insurance, makes up a far greater percentage of our national expenditure (measured against GDP) than in any other country in the Organisation for Economic Co-operation and Development (OECD). Yet the return on that investment in our overall health is embarrassingly low. Our health outcomes are considerably worse than those of our peer countries. We have lower life expectancy and higher infant mortality. On the latter statistic, we come in nearly last among developed countries! We are one of only two countries (the other is the Dominican Republic) with rising rates of maternal mortality, which is occurring primarily among Black mothers.[25] More American babies are born with low birth weight than in our peer countries. Americans suffer more injuries and homicides, adolescent pregnancy, sexually transmitted diseases, HIV/AIDS, chronic lung disease, and disability than those living in other industrialized countries. And on nearly every front, there are alarming health disparities by race, a fact the COVID-19 pandemic has laid bare.[26]

It's not that we get nothing for our healthcare money. We are tops at speed and quantity of medical services.[27] It is possible to get the best medical care in the world in the United States, and to have it covered by insurance, too, especially for problems demanding very expensive, high-tech interventions. For example, I never have a problem getting insurance to cover a child's cochlear implant surgery. But it is much harder to get coverage for other kinds of care such as the speech therapy that is needed after the cochlear implant surgery to maximize its effectiveness.

Over the last decade, a revelatory new idea has been added to the conversation: Direct spending on healthcare is far from the only spending that determines how healthy a population is. Most of what affects health happens outside the clinic. Some 40 to 90 percent of health outcomes can be attributed to such factors as nutrition and food insecurity, environment and exposure to toxins, and level of social integration or isolation. If those factors are considered part of the equation, the relative spending of all OECD countries adds up differently. The United States spends far more on medical care than most of the other developed nations do. But for every dollar that those nations spend on medical care, they spend two dollars on the social services that address social and environmental factors and promote healthy living. By contrast, for every dollar *we* spend on medical care, we spend only 90 cents on social services.[28] That's an eye-opening difference, and it helps to explain why our health outcomes are so comparatively poor despite the amount of money we spend on medical care.

These vital nonmedical factors are known as social determinants of health. The World Health Organization defines them broadly as "the conditions in which people are born, grow, live, work and age."[29] In other words, they make up the external torrent that carries us through life—the same circumstances that affect a child's level of toxic stress—and they encompass everything from the walkability of neighborhoods to the distance to the nearest hospital. They include exposure to lead and air pollution, which is strongly linked to asthma, and exposure to violence and substance abuse.

These social determinants often go hand in hand with socioeconomic status because income, wealth, and education have so much to do with the circumstances we are born into. And they affect our biology in many ways, right down to the way our genes are expressed. More immediately, they affect who becomes sick or injured in the first place. Not having enough to eat, exposure to gun violence, or growing up in a homeless shelter all have a negative impact on children's health. And other less extreme factors do, too. For instance, a low-income hourly worker like Randy who is docked $100 for taking time to go to the doctor is more likely to work when ill and, therefore, more likely to be sicker for longer and perhaps get worse, and more likely to spread disease to others.

Randy's children are also more likely to suffer the consequences of a prolonged illness that prevents him from giving them the care and attention they need—and that he wants to give them. In this way, the things that make up the social determinants of health can trigger a chain reaction that threatens children's development. Parents who are sick or struggling to put food on the table will almost inevitably engage in fewer serve-and-return interactions with their children, which reduces language exposure, which lessens the connections that young brains are making—their wiring.

We must recognize that development is the product of what physician Andrew Garner calls "an ongoing, dynamic, but cumulative dance between nurture . . . and nature." Thinking about it this way forces a shift in how we think of healthcare. As Garner writes, "Health is a continuum between disease and wellness, and early experiences play a pivotal role because the foundations for both disease and wellness are built over time."[30]

It is no mystery, then, what children and families need. Attention must be paid to building brain health from day one, and to all the ways social determinants of health shape a child's health and education outcomes. A healthcare system that does both of those things for very young children and their families would be a true safety net. It would instill confidence in all (who doesn't feel surer of themselves when they know they have backup?) and catch those who fall.

Fortunately, I've found some powerful examples of how this could work.

From the Very Beginning

By the time she became a parent, Rachel was no stranger to childbirth. She had worked as a doula, had a master's degree in maternal and child health, and was certified as a lactation consultant. But none of that was the same as having her own baby, and Rachel was anxious. "Even with all that knowledge and experience, there were things about becoming a parent that I wasn't prepared for," Rachel said.[31]

Rachel also knew that when she brought baby Eleanor home from the hospital, she would be on her own much of the time. Her husband, James, got no paternity leave and they lived in Durham, North Carolina, far away from both their families. Rachel worried about becoming isolated.

Given that every parent begins this journey as a novice, even someone like Rachel, they all benefit from a chance to check in with an expert. In many other countries, they get exactly that as well as what can only be seen as a cornucopia of support around the birth of a baby. In Finland, new parents go home with a baby box filled with about fifty necessary items like diapers and onesies (the box doubles as a bassinet if needed). In Malaysia, most mothers combine modern medical care with a traditional period of confinement after giving birth called *pantang*, during which they enjoy spiritual and social support from a midwife and follow a series of nurturing rituals to help speed recovery.[32] In the Netherlands, during the first eight days of a child's life, the *kraamzorg* service provides every mother with up to forty-nine hours of nursing care at home. The nurses can help with a home birth, and with breastfeeding or the ins and outs of formula (whichever a mother prefers). They look after both mother and baby's health, and they even help with light housework![33]

Here in the United States, most of us aren't so lucky. But that is beginning to change. A few states have adopted the Finnish baby box practice, and there are now some effective universal home visiting programs like Family Connects, which is currently operating at sites in thirteen states, with plans to expand its reach elsewhere.[34] Family Connects sends experienced nurses to the homes of new parents from late in pregnancy through the first twelve weeks of a baby's life. Randomized clinical trials of the program have found multiple benefits. For instance, in one study, mothers were about 30 percent less likely to experience postpartum anxiety, bonding between new mothers and infants was strengthened, and the parenting skills of new mothers and fathers improved. In a follow-up study when the children were five years old, participant families had 39 percent fewer referrals for child abuse investigations, and 33 percent less total use of emergency medical care for the children.[35]

When Rachel and James heard about Family Connects, they were eager to participate. The nurse arrived at their home for the first time about three weeks after Eleanor was born. In the blur of those early sleepless weeks, Rachel was a little overwhelmed. Having the nurse come to her home eased her mind. It also freed Rachel from the daunting task of taking the baby out in the winter, bundling her up against the cold and braving potentially icy and treacherous roads. Rachel didn't even have to get dressed! She was still in her pajamas when the nurse arrived for that first visit. Eleanor had been born very small and it was a relief to have the nurse check the baby's weight and health and help Rachel with breastfeeding challenges. Rachel liked that the nurse listened to what she was feeling, responded to what she needed, and acknowledged the huge changes motherhood brings. "When a baby is born, it's really not just the baby that's born," Rachel said. "A new mom and a new family are born, too."

What we need to know as parents doesn't end with diapering and breastfeeding. The next step in this new approach to early childhood healthcare is to make the pediatrician's office a consistent source of affirmation and encouragement, of information on all aspects of parenting, but especially on the essential engagement between parent and child. Communication about strengthening and reinforcing parental skills for optimal brain building *can* be woven into every layer of the pediatric office visit, from the moment a parent and child enter the front door. I've seen this in action with Reach Out and Read, a wonderful program that uses books as a tool to promote early language and strengthen parent-child relationships. Reach Out and Read incorporates shared parent-child reading times in well-child visits. Founded in 1989, the organization now has more than six thousand participating sites across the country (primarily clinics serving low-income families) and reaches over 4.5 million children a year. Studies of Reach Out and Read have shown that children who participate have significantly higher receptive and expressive language scores than children who don't.[36] And a 2021 survey of more than 100,000 parents found that those who were participating in Reach Out and Read were 27 percent more likely to read with their children than parents who weren't participating. (That result echoes earlier studies.) More to the point, participants were also significantly more likely to use reading strategies that promote warm and engaging parent-child interactions, such as talking about what is happening in the pictures and asking the child what he thinks will happen next.[37] "We're giving parents a tool and some guidance, to build on the way that their children love to hear their voices and love to be with them," says my friend Perri Klass, national medical director of the program, and a strong voice in pediatrics. "From the very beginning, small children are growing up with books and a love of reading, because they are meeting the printed word through the medium of that beloved voice."

When families arrive for their regular visits, the receptionist greets children by saying, "You're going to get a book today!" (Before the pandemic, there were shared books in the waiting room and a cozy nook for reading them.) Inviting posters on the walls celebrate books. As soon as the pediatrician enters the clinic room, she hands the baby a developmentally, linguistically, and culturally appropriate book, which the family will be able to take with them when they leave. Watching the baby manipulate the pages of a board book helps demonstrate to the doctor what he is capable of developmentally, and it distracts the baby while his ears are examined and his belly poked. The wonder of words and what they can do for the brain are laced through every minute of the visit. When the doctor measures the baby's head circumference, for example, she might joke that "this is why all the talking and reading to the baby is so crucial; it's like food for the baby's brain, helping it grow." And in a lovely example of the way pediatricians can and do extend their role outside the clinic, Reach Out and Read also partners with local libraries (one of my favorite Reach Out and Read pediatricians, Dipesh Navsaria, even has a graduate degree in children's librarianship!). Why stop with one book when children can be exposed to a library full of them?

As a hub, the pediatric clinic can play an even more important role as a link to crucial services to address the social determinants of health. "High-quality team-based care is the way to leverage the amazing pediatric platform to reduce health inequities," Rahil Briggs of HealthySteps told me. HealthySteps provides the kind of tiered, universal help that can fill the gaps—for instance, by making sure that healthy brain development is discussed early and often. The program now reaches more than 350,000 children across the country, and studies show it has positive impacts on children, parents, and pediatricians.[38]

These positive results hinge on the powerful relationship families establish with HealthySteps specialists, one of whom Rahil introduced me to. Deyanira Hernandez, a HealthySteps community health specialist, is based at Montefiore Hospital in the Bronx, New York, which is at the forefront of the changing world of pediatrics.[39] When she was a teenager in the Dominican Republic, a hurricane wiped out many homes in Deyanira's community and she spent weeks working through her school to help families in need. The experience inspired her to make helping others her life's work, and that is what she does every day. "We are there with a family from day one," she says, "throughout every developmental stage of the child." The goal is to establish a relationship, and that's just what Deyanira did with a woman named Anna.

The night Anna spent on a park bench with her four-year-old twins was one of the bleakest nights of her life. For the previous six months, she and the boys had been living with relatives in the Bronx. They slept on the living room couch, which was crowded and uncomfortable, but at least they had a roof over their heads. One of her sons had been becoming unruly at times, especially when the apartment got noisy, which was often. The relatives lost patience with his behavior and kicked Anna and the kids out. With no one and nowhere to turn to, Anna went to a nearby park, and that's where they slept.

The next day, Anna had a well-child appointment for the boys at a pediatric clinic at Montefiore. Luckily, Montefiore's doctors know to ask questions like: "Is there anything else going on? Is there anything that you need help with?" They routinely screen for adverse childhood experiences (ACEs) and social determinants of health.

Anna tearfully admitted to the pediatrician that they had spent the night outdoors and then she asked for help. The pediatrician immediately called Deyanira.

"We need you," he said.

When Deyanira got to the clinic where Anna was waiting, the young mother's relief was palpable.

"This was the only place I knew that was safe," Anna told Deyanira.

"She was in a very tough situation," Deyanira says. "It was heartbreaking."

After the pediatrician had seen the children, Deyanira took charge. In a cab paid for by the clinic, she took Anna and the boys to a nearby emergency shelter where the trio could stay the night. It was nearly 9:00 P.M. by the time Deyanira went home that night. And she met up with Anna again the next morning. Over the next few months, she worked with the shelter to find stable housing for Anna and the boys. After six months, the family was assigned to a small apartment. I suspect that if Sabrina had had an advocate like Deyanira, she wouldn't have been stuck in a shelter for more than two years.

But Deyanira's dedication didn't end there. The pediatrician at Montefiore recognized that Anna's son who acted out had undiagnosed autism, which explained his unruly behavior in the noisy apartment. So Deyanira got the little boy into therapy, and she educated Anna about how to work with her child. A few years have passed, and Deyanira still checks in regularly with Anna even though the boys, at eight, have aged out of her care. "Both children are very stable," Deyanira told me. "The boy on the spectrum has ABA therapy, occupational therapy, speech therapy. He was nonverbal and now he can say whole sentences." In their last conversation, Anna told Deyanira, "If it weren't for you and the hospital, I don't know where I'd be!"

Rahil says the work of HealthySteps is about building trust. "If we do it right, parents will start to think of the pediatrician not just as the place to fill out my school form and make sure my kid gets his shots, but also as the place where I'm asked about my social determinants of health, where I'm asked about my own depression or anxiety." As it did for Anna, HealthySteps refers patients to educational programs and early intervention. It addresses food insecurity by supplying bags of food and providing lists of food pantries. And it can get legal help for immigrants and others who need it.

There's a similar successful program called DULCE (Developmental Understanding and Legal Collaboration for Everyone), which so far exists in thirteen pediatric clinics across three states. Like HealthySteps, DULCE adds a family specialist to every team, someone who works at the nexus of early childhood, health, and public interest law and who serves all families at a clinic, not just the families with lower incomes or bigger problems. In a randomized controlled trial, researchers found that families referred to DULCE accessed supports at roughly twice the pace of other families, they were more likely to complete well-child visits and immunizations, and they were less likely to use emergency room care.[40]

Comprehensive, Connected, Cohesive

I imagined what might have happened if Keyonna and Michael had had access to the legal services offered by DULCE. How might things have been different if a DULCE specialist had leapt in to connect Keyonna with the legal support she needed for Michael, the medical (and psychological) support she needed to help her son

Cash manage his sickle cell anemia, and the information on stimulating brain development she needed for Mikeyon, who was born
while Michael was in jail? Such a specialist would have seen clearly
that Michael's arrest, Cash's health, and the new baby's optimal development were connected, and that even small improvements or
setbacks in any one of these parts of Keyonna's life would have an
effect on the others.

As Keyonna's circumstances make clear, this reimagined world
of healthcare covers a lot of ground. That's why it's so important that
healthcare providers partner with other organizations. This isn't
about putting more on the shoulders of providers. It's about knitting
together our existing patchwork of programs and leaving nothing
out. Recognizing that we need an integrated system is the last step
in putting the pieces together and making truly holistic healthcare
a workable reality. In Tulsa, Oklahoma, and Guilford County, North
Carolina, communities are building systems just like that.

In Tulsa, the Birth through Eight Strategy for Tulsa (BEST) works
shoulder to shoulder with families to make it easier for them to access services from preconception through the early years.[41] Nonprofit
and public agencies across the health, education, social service, and
criminal justice sectors have joined forces to advance the shared goal
of increasing opportunities for all children. Together, they are reaching families where they are—such as in doctors' offices, churches,
homes, and schools—and they are spreading the word about healthy
brain development everywhere they work. Thanks to BEST, every
child born in a Tulsa County hospital receives a visit from a specialized nurse who shares information about the importance of early
brain development, and how to engage in safe and nurturing interactions with an infant. In the hospital with the greatest percentage
of high poverty births, each newborn's family has access to a Family
Connects home visit. Over 60 percent of the mothers who are offered
a visit complete the program.

In Guilford County, people are thinking big as well. After years of uncoordinated investing by both government and private agencies, children in some programs were benefiting, but overall, children's outcomes were not improving across the county—only about half the children were arriving in kindergarten ready to learn. Together, parent groups, doctors, early childhood providers, businesspeople, and elected officials created something new, a coordinating central nonprofit organization called Ready for School, Ready for Life. Known locally as Ready, Ready, the new organization serves as a "backbone" and encompasses the efforts of four existing programs, each with a proven track record of success and standing connection to the healthcare system, including three of those described above—Family Connects, HealthySteps, Reach Out and Read—in addition to Nurse-Family Partnership, which sends a nurse to first-time parents with higher needs during the first years of life. (In the next phase of its effort, Ready, Ready plans to extend the continuum by working with the school system to serve older children.)[42]

In Guilford County, they are betting that the secret sauce is a program that is both deeply collaborative and universal, serving every single family in the county in a tiered fashion according to risk and needs. "It takes a total commitment by everyone to change an entire system from individual programs doing good work in silos to community-wide collaboration to multiply impact," says Natalie Tackitt, the North Carolina coordinator for HealthySteps. (Her excitement was palpable during our conversation: "It's an amazing time to be in Guilford County!" she couldn't help adding.) Ready, Ready provides a team of "navigators" to help guide families along their journey—connecting them to the right thing at the right time and linking the work of the various organizations to avoid redundancy and to create a cohesive system.

Already, the universal approach has reminded all involved just how much *everyone* needs support. When Family Connects expanded from serving only families on Medicaid to all residents in Guilford County, participation among eligible families nearly doubled, to more than 80 percent of families, according to Natalie. And pediatricians who assumed their patients didn't need anything that HealthySteps had to offer were surprised to discover how wrong they were. HealthySteps specialists ask some different questions and phrase things differently, too—they're focused on the parent, not just the baby. "Frequently, Mom opens up and bursts out crying," Natalie says. When the specialist reports back to the doctors, pediatricians are often perplexed.

"But she was fine," they say.

"No," says the specialist, "she was just giving you her company manners."

The heart of what they do, Natalie says, is attend to the critical relationship between caregiver and child. "We're nurturing the parent so the parent can nurture their baby."

Revolutionary Common Sense

Out of such simple statements can come radical changes. Healthcare revolutions usually mean a new drug or a new technology (like a cochlear implant or a new vaccine). But the reimagination of healthcare I've described will be decidedly low-tech. There is no need for flashy solutions when the answers have been right in front of us all along, as they were for Peter Fleming and the parents of Bristol. In this quietly revolutionary way, the kind of integrated, collaborative healthcare being practiced in Tulsa and in Guilford County can become a trusted voice guiding moms and dads in their parenting journey.

To maximize the long-term effects of all these early interventions, we must make brain development our North Star. Everything that neuroscience has shown us about the developing brain—its trick of plasticity and the way it is shaped by environment—means that the old way of doing things just doesn't make sense anymore. When we fully recognize all the factors that impact brain development and health, we shift from asking, *What is wrong with this patient?* to asking, *What has happened to this patient?* and *What might happen in the future?*

An ounce of prevention, as the saying goes, is worth a pound of cure. Health affects brain development in the short term and in the long term. A strong body of evidence links early cognitive and behavioral development to some of the most prevalent health issues we see in adults, such as cardiovascular disease and stroke, hypertension, diabetes, and obesity. The long-lasting health benefits from an engaging early childhood education are powerful—positive impacts show up as many as forty years later. All of this means that addressing the whole of children's lives early on costs society less in the long run. We are going to pay one way or another, says Natalie Tackitt. "So we can pay less now and people can have a better, happier, more productive and nurtured life. Or we can pay later to try and repair the damage or deal with the fallout."

Peter Fleming made a dramatic change in children's health—a 90 percent decrease in deaths from SIDS—by listening to parents, by collaborating with others, and by being open-minded. Just imagine the possibilities if we put those lessons to work again and make healthcare an anchor of our parent nation.

TEN

THE BUSINESS OF BUSINESS IS . . .

"Goodness is the only investment that never fails."

—HENRY DAVID THOREAU[1]

The memorial service for my husband, Don Liu, was a solemn event. The tragedy of his death didn't devastate only me and the children, it sent our whole community reeling. Don was chief of pediatric surgery at the University of Chicago Comer Children's Hospital, a hospital he helped found. He was not just respected, he was beloved. No one could believe he was gone. When we came together to remember him under the great arching vault of the University of Chicago's Rockefeller Chapel, sadness and shock were written on the faces of the hundreds of people packed into the pews.

Then Chris Speaker got up to remember Don. For ten years, Chris worked by Don's side as his surgical nurse.

"Don promised everyone everything in the world," Chris said. "It was my job to make sure it worked out." That meant keeping Don's surgical schedule moving efficiently, preparing patients for surgery, and then assisting Don in the OR just like my A-team helps me.

"Don always found the time, no matter what, to do the biggest cases whenever he was called upon and he did those cases with mastery that put me and everyone else in awe," Chris said. "He always adjusted his schedule or went the extra mile for any patient and family so that they were happy."

But there was a piece of Don's schedule that was known only to Chris. It started one Friday when they had six surgeries to get through. Don told Chris it was going to be especially important to finish on time that day. "I have a five o'clock meeting with Jeff Matthews that I can't miss, so call the OR and make sure they're ready early," Don said. (Dr. Matthews was the chair of surgery.) The day proceeded without a hitch until the last case threatened the tightly orchestrated schedule. It was a premature baby in the NICU who had inguinal hernias. There was a delay bringing the baby to the OR because the necessary staff was not available. The solution? Don—"no task too small"—said he and Chris would transport the baby themselves, which is definitely not in the usual surgeon-in-chief job description. They collected the baby, and the surgery went smoothly.

By 4:50 that Friday afternoon, Don was ready to leave the operating room. "You were great today," he told Chris. "Gotta go." Perhaps forty-five minutes later, after overseeing the baby's safe return to the NICU, Chris left the hospital for his commute home. His walk to the train took him through the campus of the University of Chicago and past a local park where a baseball game was being played.

"I could hear cheering and the soft, distant smack of the ball hitting the catcher's mitt, and the umpire yelling 'Strike!'" he remembered.

As he looked over at the sidelines of the baseball diamond, Chris noticed a tall, dark-haired figure in green scrubs who looked an awful lot like Don. But no, he thought, Don is in his meeting with

Dr. Matthews. Then he looked closer. It *was* Don. He looked closer still and recognized someone else. The strikes were being thrown by our ten-year-old son, Asher.

"The Jeff Matthews meeting was, in fact, Asher's baseball game."

At that, the crowd packed into the pews at Rockefeller Chapel erupted in laughter.

Then Chris added a coda. The following Monday morning, he went to Don's office and found him with his feet up on the desk, reading a medical journal.

"How'd the meeting with Jeff go?" Chris asked.

"It was great," Don answered. "Thanks for your help on Friday."

"No problem . . . by the way . . . how many strikeouts did Asher have?"

Don gave a mischievous confession of a smile. In a soft voice and with a hint of delight and pride, he said, "Eight."

For the rest of that baseball season, Chris and Don used "meetings with Dr. Matthews" as code for Asher's baseball games. And Don rarely missed one.

The story as Chris told it made us laugh. But looking back, I see it in a different light. No one who knew Don would ever have questioned his commitment to his work—to caring for children, to his patients, colleagues, and students. And everyone in that chapel knew he was equally devoted to me and the kids. His passion for baseball was another given, as was his pride in Asher's budding prowess as a Little League pitcher. Why, then, did Don think he couldn't be honest with Chris about going to Asher's game? After all, he had started work at 6:30 in the morning that day and performed six surgeries. He was hardly shirking. Yet even Don found it necessary to hide his desire to watch his son play. This is a prime example of what Brown University economist Emily Oster calls "secret parenting."[2] It's the feeling that while at work, we must pretend our children don't exist, that they have no call on our time or our attention as parents. Almost all of us have known that feeling. Women feel it more than men, but if men were immune, Don wouldn't have pretended to be at a meeting.

He must have worried that going to Asher's baseball game at 5:00 on a Friday evening might make him look like a less than ideal worker, like he wasn't sufficiently dedicated. In service to that ideal, we have erected an unbreachable wall between our parenting lives and professional lives and we behave as if there are no baseball games, no children with fevers, no scheduling conflicts, no babysitters who cancel. But the wall is a facade. All those things do exist and sometimes they affect our ability to work and our ability to parent. We would be better workers, better role models, and better parents if we no longer had to pretend they didn't.

COVID-19 made abundantly clear the extent to which our private and professional lives intersect. Our lives are porous, and our work lives don't have to be walled off from our private lives. If parents are to fully engage in their role as brain architects and loving caregivers, which is in all of society's interest, our work lives *can't* be entirely walled off from our private lives. It is time for businesses and employers to embrace their citizenship in our parent nation. This was true before, but it is even more urgent in the wake of the earthquake that the pandemic triggered in our work and personal lives. That embrace begins with a shift in culture and values, which must lead to a shift in behavior through policies that allow employees to be both workers and parents.

What Is the Business of Business?

How did we get to this place, where mothers and fathers are expected to parent both "intensively" and "secretly"? In large part, we are here as a consequence of our country's prioritization of profits and the sidelining of people. The same economic forces that drive the need for intensive parenting—increasing both inequality and higher returns on educational investment—also give rise to the need to be quiet about it.

In a striking non-coincidence, this phenomenon gained real currency (pun intended) about fifty years ago. That moment in time turns out to have been a particularly pivotal one. In 1970, just one year before Nixon's veto of the CCDA, University of Chicago economist Milton Friedman wrote an enormously influential essay in *The New York Times* entitled "The Social Responsibility of Business Is to Increase Its Profits."[3] Friedman argued that the *business of business was business* and that any social or environmental concerns that did not deal with returning more profit to shareholders had no place in the work world. In this view, employees exist purely to make more money for their companies. It follows that whether they are parents is of no interest or import to employers, although Friedman didn't even bother to mention families in his essay. He was writing at the tail end of the era of male breadwinners at work and female caregivers at home. About 40 percent of women were in the workforce by that point and the worst ravages of the decline in manufacturing were still to come. Friedman did discuss at length the idea that "only people can have responsibilities" and that any kind of social or environmental responsibility could fall only to individuals—not corporations. He suggested personal responsibility as a solution to social problems, an idea that paralleled the historic celebration of American individualism and had the advantage of letting employers off the hook.

That kind of thinking perpetuated the vision of the "ideal worker," someone who begins work as a young adult and carries on full-time for a good forty years, focusing all that time solely on work and leaving domestic concerns at home.[4] The ideal existed long before Friedman—it dates to the industrial revolution—but the Friedman doctrine's singular focus on profits reinforced the idea and made it a permanent fixture of society. Most of us have internalized it, just as Don did.

That doesn't mean we like it. Jade, who worked at Starbucks for so many years, is now a legal assistant working nine to five, Monday through Friday. She took the job for the higher pay it offered, but it was an adjustment after the flexibility she had had at Starbucks. She told me she finds it difficult to cram everything that needs to be done into the evenings—cooking dinner, helping with homework, and otherwise keeping her kids on track. Jade wishes her new bosses would show compassion for the juggling that parents must pull off. During reviews, she'd love to hear "You're human" or "Was there a lot going on?" But those words are never spoken. "You just have to pretend everything's fine and keep doing your job. There's not a lot of leeway for a woman in the workplace," she says. "I feel like society says there is, but no."

Of course, the impact of the Friedman manifesto extended well beyond enshrining "ideal workers" and "secret parenting." There has been a steady erosion of worker protections and a worrying increase in economic insecurity for individuals, leaving them ever more at the mercy of unexpected and disruptive events. For much of the twentieth century, in the United States, employers were the main source of health insurance and retirement pensions, with Social Security, Medicare, and Medicaid serving only as backstops. Since the 1970s, however, in what political scientist Jacob Hacker calls the "Great Risk Shift," many of the economic burdens and risks that were once borne by institutions have been loaded onto the shoulders of individuals and families.[5] As a result, unexpected illness or sudden job loss can easily send a family spiraling into debt or cause them to lose their home. This is what happened to Sabrina's family, and to so many others—a reality that came more clearly into focus during the pandemic, when millions of families who fell behind on rent faced eviction.[6]

The incomes of average American households now rise and fall with alarming regularity—their volatility nearly doubled from the 1970s to the early 2010s. Personal bankruptcy is also far more common. So are home foreclosures, which increased from one in 300 households in the 1970s to one in 20 in the 2010s. Should they lose their jobs, about 70 percent of Americans don't have enough savings to go more than six months without financial hardship.[7] In all, more than 75 percent of American households are "financially fragile," according to the Pew Charitable Trusts.[8] This piling on of risk affects vast numbers of people across the socioeconomic spectrum, from the poor and working class to educated professionals. Hacker writes that rising insecurity "has altered and sometimes dashed the most fundamental expectations associated with the American Dream: a stable middle-class income, an affordable place to live, a guaranteed pension, good health insurance coverage, greater economic security for one's kids."[9]

Epicenter of Risk

Almost every family I've interacted with on my journey has a story that illustrates the painful effects of an economy focused on near-term shareholder profits and not on the ways investments in children and families enhance long-term profits. When he was paving parking lots, Randy was part of the gig economy, taking side jobs and working nights and weekends to bring home whatever extra money he could. Talia Berkowitz had to leave the workforce because the high cost of childcare wasn't feasible on her income. Mariah couldn't earn a living wage while working in childcare even though she loved the job and excelled at working with kids with special needs. Kimberly Montez had no paid family leave when her daughter was in the NICU. Although some of the parents solved their problems on their own (Randy got a new job and Gabby took on extra teaching to pay for childcare), such individual fixes don't bring about real change.

Having a family has become such a frightening gamble that more and more young Americans are just saying no. They are putting off parenthood until later in life or choosing not to have children at all. The birth rate—the number of babies born per thousand women between the ages of fifteen and forty-four—continued to decline in 2020 and has fallen about 19 percent since its recent peak in 2007.[10] Most of those young people look at today's working world and don't see a way to be both an ideal parent and an "ideal worker." "The family used to be a refuge from risk," says Hacker. "Today, it is the epicenter of risk."[11]

Our system was built for a different era. The old norm—of two-parent families with a breadwinner and a caregiver—is no longer realistic or tenable. That was predictable, given the loss of worker protections, the entry of so many more women into the workforce over the last fifty years, and the lack of affordable childcare. The percentage of dual-earner households more than doubled between 1960 and 2000 from 25 percent to 60 percent—a change driven largely by economic necessity.[12] More than 30 percent of households are headed by single parents.[13] And as parents struggle to balance work and family, employers are losing out. Up to five million more workers would join the U.S. workforce if American businesses offered more family-friendly policies, according to the Federal Reserve Bank of San Francisco. (The same report shows that Canada, which has instituted parental leave and childcare subsidies, has maintained a higher workforce participation rate, especially among women.)[14] This is of national significance because the gross domestic product (GDP), the total value of all goods and services produced, is driven by growth in the workforce and growth in productivity.

Even so, most workplaces still carry an anti-parent bias (for both mothers and fathers), and young people recognize it. Women fear the motherhood penalty—childbirth is associated with earnings losses of

20 to 60 percent.[15] Seventy-five percent of American mothers have passed up jobs or promotions or have quit to take care of their kids, according to a 2015 poll conducted by *The Washington Post*. And men are no longer exempt—it used to be that having a family increased men's financial stability and earnings, but that is no longer the case. The same poll found that 50 percent of men say they, too, have had to give up professional opportunities because of childcare issues.[16] We saw this in stark relief during the Great Resignation of the COVID era when people left work in droves. As with parenting, workers too often feel the strain as an individual failing. They feel guilty when they can't make it work, like Talia, who felt she was "failing in all parts" of her life.

Among my own colleagues, I see how we internalize these ideals. Women in medicine are especially squeezed, even though, as of 2017, they make up a slight majority of medical school students.[17] I know from experience that becoming a doctor is a punishing process, requiring years of training and long hours. Like most female residents, I delayed having children because of the pressures of the work. Of those residents who have babies, some take only two weeks of maternity leave. They just don't feel they can take any more time without harming their careers or burdening their colleagues. In one study, more than 80 percent of ob-gyn program directors thought that becoming parents diminished the performance of their trainees. (These are senior physicians whose job is to bring babies into the world!)[18] A 2019 study showed that, within six years of ending their training, almost 40 percent of female physicians were either working only part-time or had left the workforce, primarily due to work-family conflict.[19] What a loss to the medical community and to society to have these women—who have invested years in training—feel that they cannot do the job and also be a parent.

At TMW, my colleagues spend their days immersed in the importance of parental engagement in the earliest years of life. And they, too, worry about taking time off to have children.

"Do you have a second to talk after our meeting?" Dani Levine asked me solemnly one day. I could feel the trepidation in her voice. Dani is a brilliant young developmental psychologist, running our SPEAK research program. I could feel my throat tighten. Throughout our meeting, the mother in me couldn't stop worrying about her. Was she sick? Was she in some kind of trouble? Finally, the meeting ended, and we were alone.

"Dana, I have some news . . . I'm going to have a baby."

I couldn't help but burst out laughing with joy.

"One more baby for the TMW family!" I cried.

Dani later admitted she was petrified to tell me she was pregnant and was so relieved by my response. Six months later, she had a beautiful baby girl with fat cheeks like my Genevieve. And now we are working out the details of a part-time return to work so she can have more time with little baby Maddy. Dani is slowly wading back into the workforce. And that is great.

On one hand, I was amazed that anyone would think I would respond to news of a new baby with anything other than joy. On the other hand, I get it. I understand why, for some organizations, maternity leaves trigger more gulps than glee. Running TMW is a lot like running a small business. Dani is the only person who does her particular job. A long maternity leave could have meant that her work came to a complete halt. In our case, we had hired a talented postdoc, Caroline, not long before Dani got pregnant and trained her to step in while Dani was away. But if announcing a pregnancy is anxiety-provoking at TMW, I can only imagine what it must be like at less welcoming organizations.

It doesn't help that we often look around and assume that others have it figured out. When my children were growing up and I was building my surgical practice and establishing TMW, people would ask: *How do you do it?* It's a common question for any woman with a high-powered job who is also raising children. The answer for me—and I imagine for most others—is that there is a lot of unseen labor that makes my life work. In short: I had help, and lots of it, not just from Don but from nannies, housekeepers, my mom, my dad, and others. None of us is a superhero with magical abilities to manage work-life balance; we are all just human. We become heroes when we admit this and acknowledge that some people have more resources to throw at the problem, while others are their family's *only* resource. As sociologist Jessica Calarco so aptly puts it, "Other countries have social safety nets. The US has women."[20]

The shift to remote work and school in the pandemic made the challenges of balancing work and family life abundantly clear, and we cannot go back to the old normal.

Making the Workplace Work for Families

There are bright spots out there, to use a favored phrase of one of my favorite friends, Ralph Smith, who runs the Campaign for Grade Level Reading. "Don't underestimate the transformative potential of a good example," Ralph says. One such example is Starbucks. The same place I turn to for my iced coffee is the place Jade turned to for health insurance and flexible hours when she couldn't be the stay-at-home mom she always imagined she'd be.

It was an early, searing experience that inspired Howard Schultz, longtime CEO of Starbucks, to do things differently. Growing up in public housing in Brooklyn, Schultz's story goes, he was the oldest child of parents who didn't finish high school. There was never enough money. One day when Schultz was about seven, he came home to find his father "distraught," lying on the living room couch in a cast. It was winter and Fred Schultz had been making deliveries for a cloth diaper service when he slipped on an icy sidewalk, breaking his ankle and his hip. He was immediately fired. He had no health insurance, no workers' compensation, and no savings.[21] Just like that, Schultz's father didn't just fall on the sidewalk, he fell out of the workforce and into instability.

"The image of my father on the couch, helpless, stuck with me," Schultz has said. He points to that day and to his childhood of struggle when he seeks to explain the motivation behind his style of leadership. "I never set out to build a global business," Schultz says. "I set out to build the kind of company that my father never had a chance to work for. One that treats all people with dignity."

Schultz made that dignity a reality with concrete policies. That is why Jade turned to Starbucks when she needed work, and why she stayed there as long as she did. She could schedule her shifts for when her husband or mother could be home with her son Nathan. And the benefits were good, even for a part-time barista. The company provided health insurance when her family couldn't get it elsewhere. Anyone who works twenty hours a week for the company is eligible. When Schultz instituted that policy in 1988, it was practically unheard of to give part-time workers health insurance; it still is. Schultz didn't stop with healthcare. Starbucks gives *every* employee stock in the company. It joined with Arizona State University to provide employees (they're known as "partners" in the company lingo) with a free online college education. Starbucks provides paid parental leave for mothers and their partners. They have also joined forces with Care.com to provide employees with access to ten days of emergency care.[22] I think about Tatiana, who had to decide between leaving her kids alone for two days or losing her job as a hotel housekeeper. What would it have meant to her to have that kind of help? (It should be noted that some employees at Starbucks, as at many other companies, are seeking additional worker protections, and the first Starbucks stores voted to unionize as this book went to press.)

The fact that Schultz had direct personal experience of the impact employers have on families undoubtedly helped him to understand why business needs to change its culture and policies. He knows that employers have a responsibility beyond profits, and he also realizes that the family-friendly policies he has put into action are actually pragmatic, because such policies can help the bottom line. As Schultz likes to point out, Starbucks provided all those benefits to employees while delivering more than a 20,000 percent return on investment for shareholders.[23] Hard for shareholders to complain about that! Moreover, things are changing. Employees are now showing increasing willingness to vote with their feet: 83 percent of millennials say they would switch jobs if another company offered better family-friendly benefits than their current employer, according to a survey by Care.com.[24] Perhaps that's why many of the country's largest employers have begun to offer paid parental leave.

The benefits of such policies ultimately extend to the companies themselves—in giving them not only a willing, committed workforce, but a high-quality one. As Natalie Tackitt points out about one motivation for the early childhood efforts in Guilford County, "If Guilford County is the best place in the world to raise a family, you're going to have employees who are happy to work for you here. And if all the children in Guilford County are ready for school when they get there, you have an employee pool down the road that will benefit your company. So, everybody should care about this." Her words are echoed by none other than the U.S. Chamber of Commerce Foundation, which notes, "A world-class workforce begins with a world-class education system. The path to that education starts with a solid foundation constructed in the first years of life."[25]

With the neuroscience of early childhood as a guide, the map for employers is clearly marked. We can apply what we know about foundational brain development in children to inform a new way of doing business. Put simply, children need three things during the early years: security (i.e., protection from toxic stress), enrichment (i.e., rich language input), and time (i.e., the opportunity for nurturing interactions with caregivers). What their parents need is the ability to provide these things for their kids.

The way to deliver on those needs is for employers to acknowledge that we are (nearly) all parents or caregivers one way or another. If as a society we believe that it's in our interest to prioritize
the healthy development of all children, workplace culture should
embody that belief. This would mean putting into action practical
policies and programs that help parents help their kids.

Just as COVID-19 busted the myth of the traditional office, we
need to bust the norm of the ideal worker. Some companies, like the
one a woman named Elise works for, have already gone down that
road and are the better off for it, as are their employees. I got to
know Elise through my friend and colleague Liz Sablich, who had
explored sharing a nanny with Elise until it turned out Elise no
longer needed a nanny.

For Elise, there was no secret parenting. Anyone who walked into
her office knew she had a baby because he was right there. She
worked in the Washington, DC, office of the National CASA/GAL
Association for Children (Court Appointed Special Advocate/
Guardian ad Litem). CASA had a policy that encouraged employees
to bring their babies in to the office during the first six months of an
infant's life. Elise's bosses know that losing Elise and having to replace her would be more costly than making space for her baby for
a few months. (The cost of replacing an employee is estimated to
equal six to nine months of salary.)[26]

For Elise, the policy meant that after her maternity leave was up,
when her son Griffin was three months old, she brought him to the
office with her for two and a half days a week. (Her husband's schedule allowed him to stay home with Griffin the other days.) At first,
Elise was nervous, worried that Griffin would make too much noise
or be disruptive. She was worried, frankly, that she would no longer
be seen as an "ideal worker." Her office was small—there were just
four of them working there when she had Griffin, and none of the
others were parents. But quickly, her colleagues became "aunties" to

Griffin and were happy to spend a few minutes looking after him if Elise had to use the restroom or had an urgent phone call. Whenever the CEO, Tara Lisa Perry, came through town, she was excited to see Griffin and bummed if she arrived on a day when the baby was with his dad. The schedule ended up suiting Elise perfectly. She reserved her most demanding work for the days when she didn't have Griffin and scheduled most of her meetings for those days as well. Days that Griffin was there, she got through more routine tasks and waited to make calls when he napped in an empty room down the hall. "Griffin had his own office," Elise says with a laugh.

Real Change

Once employers fully accept, as Elise's did, that workers have children, the next step is to acknowledge what that necessitates. As we saw with Randy, stretched finances stress even the strongest of families. To be able to provide an enriching environment for their children, parents need economic stability and financial health—in other words, a sturdy boat to ferry their children. They can have that only if they receive a living wage, enough to be able to cover the basics of life, including housing, food, and childcare. In 2019, fifty-three million people, or 44 percent of the workforce, were earning less than $16.03 an hour (as a national threshold) and were considered low-wage workers. It's a group that is disproportionately female and Black.[27] Women make up more than 70 percent of the workforce in six of the top ten low-wage jobs (e.g., childcare workers, 95 percent of them women, earned an average of $9.48 an hour in 2014, according to Bureau of Labor Statistics figures).[28] Like so many of the families I've worked with, Randy found that working two or even three jobs was the only way to provide for his family.

While doing that, it was nearly impossible to summon the time and energy necessary to parent with full engagement. When he finally got his job at the YMCA, he was thrilled to be earning more, to have retirement benefits, and to be able to give up his extra jobs. Most significantly, Randy was thrilled to get more time—time to be with his children, time to steer the boat.

There is plenty of research showing that paying parents a living wage, and therefore giving them some stability and security, is good for kids—it reduces behavioral problems and strengthens cognitive skills.[29] The irony is that although many corporations today equate competitive wages with low wages, the original tough-minded American businessman, Henry Ford, showed that higher wages could be good for business, too. In 1914, Ford more than doubled the pay of his factory workers to $5 per day.[30] Ford did not take that step out of the goodness of his heart (he was not a nice guy and was a vocal anti-Semite). He did it to reduce soaring turnover rates and the costs of training new workers. The move was enormously successful, improving the stability and quality of Ford's workforce (while bringing many of them into the middle class). Ford considered the pay raise the best cost-cutting move he ever made!

Parents also need the time and bandwidth to be brain architects. Too few companies today understand that policies that give parents the time they need are also—like paying them a living wage—good for business. For many parents, time equals flexibility, something few workers have. In a 2018 study (pre-pandemic), 96 percent of employees said they needed the ability to change things up—to change their hours, the location where they work, the frequency of travel, and their freedom to step away briefly during the workday to meet personal obligations. But less than half reported having that kind of flexibility.[31] Without it, caregiving becomes a burden, health and wellness suffer, and employees become miserable and resentful. The result, for business, is that employees are less productive at work, more likely to leave their job, report lower levels of engagement, and have more negative views about their employer (and are less likely to be advocates of its products and services).

Flexibility is exactly what Sabrina didn't have and it's exactly what could have prevented her from living with her two children in a shelter for two years. Before she quit her job, which eventually led to her becoming homeless, the first thing Sabrina did was ask to change her hours. She thought if she could start the workday just a little later so that she could be at home in the morning to tend to her older son and help her husband learn to manage his newly diagnosed diabetes, things might be okay. But her employer said no, and because of this lack of flexibility, the first domino fell in the sequence that ended in that dismal shelter.

Unfortunately, many industries now offer the wrong kind of flexibility—"just-in-time" scheduling, in which employees' schedules change on short notice, depending on how much work there is to be done.[32] Such last-minute scheduling is all about suiting the needs of employers, not employees. For hourly workers, shifts are scheduled week to week, based on how busy any one store is, and the workers never know how long they'll be on the job, or when. One worker at a Philadelphia-area Target was promised thirty to thirty-five hours per week when she began part-time work. But soon, her hours dwindled and varied wildly—from eight hours one week to twenty the next, then twelve.[33] When that is how you must work, it is simply not possible to count on a steady income. And how exactly are you supposed to organize childcare? The irony is that such last-minute scheduling hurts employers more than they realize. In 2018, a careful study of Gap stores found that giving workers stable schedules increased both sales and productivity.[34]

Salaried workers have their own challenges. For them the problem is often that the demands of the workplace never stop. They are expected to work long hours, respond to email around the clock, and never let family conflicts get in the way.[35] It's not surprising, then, that flexibility in working hours and the ability to telecommute part of the time were prized commodities before the pandemic—they provide a little bit of control. Surveys found that people thought the ability to set their own schedule was worth giving up a 9 percent increase in wages and that telecommuting was worth 4 percent.[36] Some companies were already instituting changes in this direction before the world went into lockdown. Large companies like the Atlanta-based finance firm Credigy set up flexible work schedules that allowed employees to come and go as they pleased so long as their work got done. It took time to get people to take advantage of the change. Tellingly, Credigy found the key was making sure that senior staff, including men and people who weren't parents, were using it, too. In other words, leaders had to signal the culture change from the top down.

In addition to fair wages and flexibility, workers need some form of protection from unfortunate events and what economists call "negative economic shocks." Paid family and medical leave is an obvious first step. It would have allowed Kimberly to be with her daughter Penelope in the NICU, and it would have been a second option for Sabrina when changing her hours proved unworkable. Studies show that where it exists, paid leave improves productivity, employee loyalty, and morale.[37]

When paid leave is taken by everyone who is eligible, it also promotes gender equality. Paternity leave strengthens the bond between fathers and their children and sets up more equitable parenting habits for the long haul. Fathers who learn early how to change diapers and do feedings are more confident parents, and fathers who take longer paternity leaves are likely to be more involved with their children for years to come. Children are also healthier when fathers are more engaged.[38] (One benefit to my children—and me!—of having a pediatric surgeon as a father was that he was a pro at diapers and took charge of the poopy ones.) What's more, when fathers take longer leave, mothers benefit in some unexpected ways. In the short term, they are less likely to suffer from postpartum depression. Over the long term, they have higher earnings and their relationships are less likely to end in divorce or separation.[39]

The advent of the gig economy makes portable benefits an even more necessary tool to help families weather whatever comes along. They are just the kind of innovative idea we need to solve the problems of today's business climate—the job market may have changed but the need for benefits has not. Portable benefits are connected to an individual rather than a particular job.[40] They're not a new idea— Social Security is a portable benefit, as are retirement savings plans that travel with an employee. The health insurance provided by the Affordable Care Act is a more recent example of a benefit that isn't dependent on employment. But the idea has generated new interest lately, given the changing dynamics of the workplace. Portable benefits would go a long way to increasing economic security, and employers can and should support them.

One way to make sure you can balance work and parenthood is to become your own boss. That was the solution a young woman with big entrepreneurial dreams turned to in Austin, Texas, twenty years ago. With her first baby on the way and aspirations of a career in fashion, she invested $500 in materials and started designing jewelry in a spare bedroom. "I wanted to have a career that allowed me to put being a mom first," she said years later.[41] After her son was born, she carried him with her to local shops to try to convince them to sell her wares and brought her friend along as a babysitter when she took her child to New York, where she attended wholesale events. By the time she opened her first shop in 2009, at the height of the financial crisis, she was a divorced single mother of two boys. It was challenging, to say the least. "My two little boys were with me all the time. I couldn't afford a lot of help." But Kendra Scott persisted, and the eponymous jewelry business she led until 2021 is now worth one billion dollars, with more than a hundred stores and some two thousand employees. (To know how popular her creations are, I just have to look at Amelie and her friends, who all own Kendra Scott necklaces.)

Given that origin story, it's not surprising that Scott, who had a third son when she remarried, has established a thoroughly "family-first" company culture. More than 90 percent of her employees are women, many of them young mothers. The company offers benefits that deliver exactly what parents need to give their children: time, enrichment, and security. It gives parental leave to everyone, including part-time employees. It offers fertility and adoption assistance. It gives generous paid time off. Scott established a family fund, supported by employee donations and by Scott herself, to help workers facing unforeseen financial hardship. It covers exactly the sort of events that trigger income volatility and insecurity, like a home damaged in a hurricane, or a family member with a sudden illness or injury. And because "life happens" in smaller ways as well, Scott's company also instituted a Pass-the-Baby policy. Babysitter sick? Childcare closed? No problem. Just bring your child to work. "I myself have sat in the circle with my co-workers on the floor for our weekly meeting with my kid playing in the middle of us," says one executive. The company says it reaps the benefits of its benefits: 95 percent of employees return after taking family leave, promotions and morale are high, and turnover is low.[42]

But change doesn't happen one business at a time. Real shifts in workplace culture and policy emerge when groups come together and believe that change is necessary. Kendra Scott's company is one of more than four hundred in Texas (small, medium, and large) that have qualified for the Best Place for Working Parents status.[43] That stamp of approval grew out of an effort by business leaders, educators, nonprofits, philanthropists, and civic leaders in Fort Worth and has since spread to other cities in the state, such as Austin, where Kendra Scott is based, as well as other cities and states nationwide. "It's not just our schools' responsibility to educate children," Sara Redington of the Miles Foundation, one of the organizers, told me. "Business leaders know education is economic development. Family-friendly practices help working parents thrive in the workplace while also supporting their children's growth and development at home."

The Best Place for Working Parents campaign lays out the top ten policies proven to benefit families *and* help businesses' bottom line. For the record, they are: company-paid health insurance, paid time off, parental leave, nursing benefits, onsite childcare, financial assistance with childcare, backup childcare, flexible hours, working remotely, and designation as a "best place" (of course). Companies can take a three-minute self-assessment to see if they qualify and where they can improve. The Best Place also gives annual innovator awards that highlight companies doing an especially creative job of helping employees balance work and life. I found similar coalitions in other states (North Carolina is one) promoting the idea in their communities that family-friendly equals business-friendly.[44]

I believe many Friedmanites would be appalled. Yet I also believe there's another way to interpret Friedman's assessment of corporate responsibility. A firm's bottom line depends on *both* near-term and long-term profits. When a farmer fertilizes his crop in the spring, it raises costs and lowers profits in that quarter, but fertilizing ensures a plentiful harvest—and profits—in the fall. Since neuroscience shows so clearly the return on investing in early childhood, a literal interpretation of Friedman's focus on profits would call for a *re*-investment in today's families, not a further *de*-investment. That re-investment will sow the seeds for the next generation of highly skilled, highly educated, and highly productive workers. The business of business is . . . investing in families and tomorrow's harvest.

At the End of the Day

Those are the economic reasons businesses should invest in children and families—they'll get happier, more productive employees today, and stronger, more effective employees tomorrow. But there is another, more powerful reason for businesses to help build a parent nation. Quite simply, it's the right thing to do. All members of a society benefit when its children receive a fair start in life. And all parents benefit from fully engaging with their children.

I don't say this lightly. Don's death was a painful reminder of what matters most in life. I know he was proud of his work. But he would be even prouder today of our three children. He missed the chance to see them grow up and that is a far greater sorrow to me—and I think it would have been to him—than the fact that his professional life was cut short.

It shouldn't take a tragedy to show us how much our children matter—not just to us but to society. They are the next generation. The humanity with which we treat parents matters, too. Business leaders have an opportunity to lead with their humanity, to treat workers with empathy and dignity.

Of course, there are constraints. Not every business can or should have babies as regular visitors. The only babies who can be in my operating room are the ones who are receiving a cochlear implant. And not every business can meet every item on the Best Place for Working Parents policy wish list. But that is not really the point. The kind of rethinking of possibilities that Elise's employer did, or that Kendra Scott did when she built a company around what she herself needed as a young mother, or that Howard Schultz did when he instituted company policies that would have saved his own childhood family from poverty, will further the cultural shift we need.

I think that if Don were alive today, this is exactly the kind of rethinking he would have done. He wouldn't have invoked imaginary meetings to be seen as an ideal worker the way he did in 2010. People like Don, respected and in a position of authority, need to lead by example. Of course, they still must put in a productive workday, but they shouldn't have to pretend that there are no kids' baseball games, no concerts, no parent-teacher conferences, that their children don't get sick, that the babysitter never cancels.

Knowing Don, I believe that, today, he would wear his parenting as proudly as he wore his scrubs. In doing that, he would help break down the remains of the crumbling wall between personal and professional lives and help build the parent nation we need. As a bonus, he would have been able to tell everyone he knew exactly how many strikeouts Asher threw.

ELEVEN

LIFE, LIBERTY, AND FULFILLING SOCIETY'S PROMISE

"It must not for a moment be forgotten that the core
of any social plan must be the child."

—FRANKLIN D. ROOSEVELT'S COMMITTEE ON
ECONOMIC SECURITY[1]

I love a good love story, and Michael and Keyonna's is one of my favorites. She was the best friend of Michael's older sister, whom everyone calls Yoyo, so the two had known each other casually for years. But Michael was six years younger, an age difference that seemed like a big deal until suddenly it didn't. As Michael tells it, their romance began on a beautiful summer day in 2012 when he spotted her hanging out on a friend's porch and looking gorgeous in a blue-and-orange tie-dyed dress and strappy denim sandals. She'd done her hair in dramatic fashion, too—dyed blond and styled spiky. When Michael kept telling Yoyo how much he liked Keyonna, she eventually said, "Mikey, don't tell me. Tell her."

On that summer day, he finally did, telling her how nice she looked in that dress. They started flirting ("conversing," as Michael says). Michael loved her confidence. Keyonna loved his sense of humor and his gentleness, and she thought he was mature for his age. Their opposite temperaments—Keyonna is "out front with whatever is on her mind" and Michael is more reserved—might have pushed them apart but instead pulled them together.

Soon they were spending all their free time together. Michael loved hanging out with Keyonna's sons, Cash and DiaMonte. Within the first year, however, the new couple faced what would turn out to be the first of many tests in their relationship. While at work, Keyonna suddenly doubled over with shooting pains in her abdomen. Frightened, she called Michael and he raced her to the hospital. It turned out she was pregnant. They were thrilled for a moment, but then the doctor explained that the pregnancy was ectopic—trapped in one of Keyonna's fallopian tubes instead of in her uterus. To avoid a life-threatening rupture, she would have to have emergency surgery to remove the tube and the nonviable pregnancy. Michael and Keyonna were crushed. They worried Keyonna wouldn't be able to get pregnant again and they would never have a child together.

In the following months, their relationship proved resilient and grew even stronger. Michael formally moved in with Keyonna and her boys. Both had good jobs (he at a meat wholesaler, she in insurance) and they were raising Cash and DiaMonte together. Their little family was happy and secure. But, oh, how they longed for a baby together. "We were just wishing on the stars," Keyonna says.

One year later, their dream came true. Keyonna was pregnant again. "We were so joyful," says Michael. "That was a big day." For those first few weeks, they would sit on the couch together in the evenings after the children had gone to bed. Michael would put his hand on her belly, and they would imagine the future their child might enjoy. They were sure it would be a girl since Keyonna's first two children were boys. Michael called the baby his "princess" and they dreamt up names that would commingle their own—Michael and Keyonna—to show the world the depth of their love.

Any birth, even when there isn't a love story behind it, represents a new beginning, a precious new life full of potential and possibility.

Birth is often a metaphor for fresh starts, not just as they apply to new life, but to momentous new endeavors. That's why we tend to speak of the birth of our country in hopeful and optimistic language. I imagine that the founders of the United States felt something akin to what new parents feel when the founders set down the framework that would bring this new nation into being. "We hold these truths to be self-evident," they wrote in the Declaration of Independence, "that all men are created equal, and that they are endowed by their Creator with certain unalienable Rights, that among these are Life, Liberty and the pursuit of Happiness."[2]

The idea that individuals are endowed with those rights—and, therefore, not beholden to a monarch—was revolutionary. It would be the basis of a country that was the first of its kind, one based on democratic principles and grand aspirations. Now, I am not so starry-eyed over the Declaration of Independence that I don't recognize that many of the conflicts and inconsistencies that trouble us today were with us then. All people were not, in fact, treated equally. Far from it. Our young nation enslaved people and it disenfranchised women. Indigenous people were killed or kicked off the land they had lived on for generations. The repercussions of those travesties echo through our history and contribute to the systemic racism that erects barriers for far too many of our citizens today.

Nonetheless, the ideas in the Declaration were stirring and ambitious. They have inspired Americans for centuries, and they inspire me today. Our belief that every citizen of this country is entitled to life, liberty, and the pursuit of happiness provides the sense of boundless opportunity that animates the American Dream. The promise of America is to grant her children the ability to fulfill their natural promise, to let them learn and grow and flourish, to allow them to become productive members of their communities and society at large.

We must recognize that foundational brain development is a prerequisite for that promise, a fundamental right of its own. It is what gives every child a genuine chance at equality. Because what are the rights to life, liberty, and the pursuit of happiness worth if you lose your shot at them within the first years of life?

A Dream Interrupted

Michael and Keyonna shared their love story with me in 2019, just two weeks after Michael was released from prison. There they were, sitting across from me, in the same third-floor walk-up apartment in North Lawndale on Chicago's West Side where they had first dreamed of their unborn child's future. So much had happened since those happy moments. But now they were once again together on that couch, his arm over her shoulder, her hand on his thigh, in that easy, affectionate way that couples have. Despite all they'd been through, I could still hear joy in their voices as they talked about the future and as they remembered the time before Michael's arrest, when they were just looking forward to the birth of their baby.

Their dream of raising their child together was interrupted on the fateful day in May 2014 when Michael was arrested for a murder he didn't commit and had nothing to do with. Because he was denied bail, he had to wait in pretrial detention until his case was heard.[3] Their child was born in late November, six months after Michael went to jail. With Michael behind bars, Keyonna's sister stayed with her in the delivery room. And Michael's two sisters arrived the following morning, Yoyo hauling a car seat and other necessary baby gear with her. The baby turned out to be a boy, but Keyonna stuck to their plan to combine his parents' names and called him Mikeyon.

She brought little Mikeyon to see Michael one month later. They met in the visitors' room of the jail, sitting in a tiny cubicle on either side of a Plexiglas divider. "She took his little hat off his head," Michael remembers. Mikeyon had been asleep, but he opened his eyes and looked at his father. "He looked just like me." Keyonna showed Michael the baby's little fingers and toes. The memory is bittersweet for Michael. He was bursting with pride, but he was distraught that he couldn't hold Keyonna or the baby. "I had missed my first child's birth," he says. It was something he could never get back, and he knew Mikeyon's babyhood was something he could never get back either.

For nearly five years, as his case crawled through the judicial system, this is how Michael, Keyonna, and Mikeyon interacted—in fifteen-minute increments, on Sundays, separated by Plexiglas. Keyonna had to pass through arduous security screenings every time she visited. Even Mikeyon's bottles of milk were scanned. When they finally got to see Michael, Mikeyon was often asleep or cranky after waiting forty-five minutes or more to get in for the visit. It helped when Cash came along, too, because he kept his little brother entertained. But, eventually, Keyonna let the boys stay home sometimes. "It hurts his dad when he doesn't want to go," Keyonna told me while Michael was still in jail. "I tell Michael, you have to understand that a visit is a long process and we have been coming here a long time."

Despite his circumstances, Michael was determined to be the best father he could. Although he had stepped in to be a father for Cash and DiaMonte, he hadn't ever had a child of his own. He had a lot to learn and eagerly signed up for a prison program for dads. "They tried to teach us how to be a good father, teaching us how to interact with our kids," he says. "I was like a sponge soaking everything up, trying to learn." Part of the incentive was a promise that the inmates who participated would get more substantial visits with their children, in person—holding their children was to be a reward. "I wanted to hug my son," Michael said. Before he and the other dads got that chance, the program was discontinued. It was a crushing disappointment. Not once in five years did Michael get to hold Mikeyon in his arms.

All that time, Keyonna had to carry on alone. "If I fall, my whole family falls," she told me just before Michael was released. She is a wonder to me. As I got to know her, I saw that her determination and fortitude had carried her through many years of challenges. She had DiaMonte when she was just sixteen, but she still finished high school, carrying her baby with her across the stage to get her diploma. While Michael was in jail, she raised the three boys with help from family and friends, including her mother and Michael's sisters. It was her mother who first saw one of our TMW posters when she took Cash in for one of his frequent blood transfusions for his sickle cell anemia. Even with all that she had going on, Keyonna signed up for our home visiting program and longitudinal study and never let on how hard things were for her. I only learned that Michael was in jail because Keyonna explained she couldn't use the LENA (the Fitbit-like device we use for recording interactions between parents and children in our studies) when they went to visit Michael on Sundays. It was too hard to get it through the security screening.

Watching her in mom mode was like watching an Olympic athlete in training. She was on it. She kept up running conversations with her kids. She tried to read to Mikeyon thirty minutes a day. "A book is like candy to him," she said. Even when she let Mikeyon watch *Sesame Street*, she sat alongside him and talked about the show so he would fully process what he heard and saw. She tried to limit her boys' technology use and hid her laptop under her bed so the kids couldn't reach it and download games that would eat up all their time. (Later, she took to changing the Wi-Fi password when she wanted them to get off their screens and engage with her.) She woke up every day at 5:30 A.M. to spend time with Mikeyon before her shift at Walgreen's, the job she took after Mikeyon was born because the hours were more flexible. In those early mornings, the two of them had some of their richest back-and-forth conversation. "From six to eight, it's all me and Mikeyon," Keyonna said at the time.

While Keyonna was holding it together at home, Michael was trying to hold on as the pain and suffering mounted in jail. During those five years, first his mother, then an uncle and an aunt died—he missed their funerals. "I lost a lot," Michael told me, his voice cracking with emotion. But Michael and Keyonna were there for each other. "She made me feel good," he remembers. "She'd say, you'll be okay, your turn will come. When I'd be down, she'd lift me back up." She even wrote him poems to keep his spirits up. Keyonna knew Michael hadn't killed anyone. "I wanted to stand by his side," she told me. "There's nothing like having a significant other just to hold you together through everything." She wanted to get him out so they could raise Mikeyon together. "We helped each other a lot through that time," Michael says. "I had to stay strong on the inside. She had to stay strong on the outside."

Finally, in July 2019, their strength was rewarded. Michael's case went to court. The DNA evidence didn't match him, the witness statements were shown to be suspect (remember one of them described the perpetrator as six or seven inches shorter than Michael), and the jury freed Michael. After five long years, it was all over in a day and a half. Upon release, the first thing Michael did was throw his arms around Mikeyon and kiss Keyonna. And then they went home to try to pick up the pieces of their dreams.

As a nation, it sometimes feels that we are picking up the pieces of our dreams, too. Or that we've just given up on them. We certainly are not living up to our ideals. We are not fulfilling the promise *of* our children nor the promise *to* them. To ensure that our natural rights begin at birth for every child, and to repair the long-standing inequities in our country, our founding principles need to be nurtured and protected just as parents nurture and protect their children.

A Chance to Get It Right

"The rain doesn't last forever," Keyonna told me when I asked her how she got through the years that Michael was incarcerated. It was the same thing she told Michael to lift his spirits. She is someone who just keeps moving, taking care of whatever needs to be done. For Michael's part, when he was finally released, after years of separation from his family for a crime he didn't commit, I know he felt anger and bitterness. Who wouldn't? But those emotions weren't his focus. Instead, he was full of the same optimism he felt when Keyonna first got pregnant. He was looking forward to new possibilities.

"I felt reborn," he told me.

Michael and Keyonna did not lose faith. They held true to what brought them together. As a nation, we should hold true to our founding principles. Addressing the needs of every child from the first day of life is the single most effective thing we can do to fulfill our promise and make good on our ideals. Strong families and children are the threads holding the fabric of society together.

The fundamental right to strong brain development should inspire us in our efforts to build a parent nation centered on children, families, and their communities. In one of my favorite quotes, neuroscientist Joan Luby of Washington University put into perspective how consequential this issue is. And she went on to show how science can inform the path forward. "Healthy human brain development represents the foundation of our civilization. Accordingly, there is perhaps nothing more important that a society must do than foster and protect the brain development of our children," she said. The science, she concluded, represents "a rare roadmap to preserving and supporting our society's most important legacy, the developing brain."[4]

Early Education as a Public Good

What neuroscience shows us is that education and learning begin on day one and that engaged and loving parents and caregivers are the key. Knowing that, we are fully equipped to plan and implement smart public policy. With brain science as our guide, it's easier to see that investments in early childhood care and education are a public good (not in the strict economic sense but in the literal sense). They are as necessary for a functioning society as public parks, roads, and fire departments, and, most relevantly, as necessary as K–12 public education.

America was ahead of the rest of the world in instituting free public education. The system's history aligns with America's history. We viewed education as the great equalizer and the key to the American Dream. And our embrace of universal public education helped propel the United States to become the dominant superpower of the twentieth century.[5] We invested in human capital.

But we haven't kept up with what science tells us children need. We don't start early enough. Although the United States was the first nation to institute free public education for ages six to eighteen, much of the rest of the world caught up and then leapfrogged ahead of us by adding, first, kindergarten and then pre-K. Almost every country that didn't have free K–12 education in 1940 had that plus preschool education by 1990, according to early childhood expert Ajay Chaudry of New York University.[6]

To build a system that includes and supports children's earliest years, we have to reimagine what we mean by universal public education. That means looking beyond what's illuminated by the streetlight—K–12 education—and understanding that such a system will need to be more than "school." It must encompass healthcare, paid leave, earned income tax credits, high-quality childcare, and much more. It's a job for a wide variety of public agencies, working together with parents, communities, healthcare providers, and businesses. Such a system can result in a new societal framework centered on children, families, and their communities, a framework that begins on the first day of life and recognizes that the child's early years are an essential part of the education and care continuum.

Fortunately, government leaders have begun to see that supporting foundational brain development is essential if they want to meet their goals of building a healthy, productive citizenry. "A lot of policymakers are more and more convinced by the science that the earliest years matter," Cynthia Osborne told me. Cynthia works at the intersection of early childhood science and policy, helping to translate the evidence from research into actionable programs and legislation. Often, she says, the thing that convinces policy makers are actual pictures of the differences in children's brains that result from differences in their life experiences—the kind of brain scanning Kim Noble does. And a recent preliminary study found that antipoverty policies do have a protective effect on children's brains, especially in states where cash benefits were more generous.[7] But even when policy makers understand what's at stake, they don't necessarily know where to begin. To help them, Cynthia established the Prenatal-to-3 Policy Impact Center at the University of Texas at Austin's LBJ School of Public Affairs. The center is a resource for state governments (because they are the ones who have to make early childhood policy work on the ground in the United States). "We aim to fill this gap and provide them with guidance on the most effective investments they can make," Cynthia says.

To my mind, filling that gap means making foundational connections in our infrastructure. Of course, the solutions in each state and community will look a little different. They must respond to the needs and desires of families, and they must be culturally relevant and responsive. Whatever the differences, however, workable solutions will all have nurturing adult-child interactions at the center. And the first thing policy makers need to understand is that investing in children's development means investing in the adults in their lives. "Care for the caregiver so they can care for the child," Cynthia says.

Like Cynthia, I have come to believe that our approach to early education and care should be a seamless continuum that begins during pregnancy and carries children and their caregivers through to the first day of school. It is what Ajay recommends in his book *Cradle to Kindergarten*.[8] It would be an amplification of the continuum we should be creating within healthcare. In this expanded approach, every program that touches a young child—whether it's early intervention, food stamps, pre-K, or something else—should be knit together as part of a coherent whole, all of it working in concert to ensure that parents can serve as brain architects to help their children forge the critical brain connections they will need.

We know a lot about what works, and we certainly know why we need to start early. People like Cynthia and Ajay have put in the painstaking hours and reviewed the evidence on hundreds of programs and policies that affect very young children directly and indirectly. Cynthia's organization arrived at a very specific list—they call it a road map, just as Joan Luby does—of the top policies and strategies that states should pursue first to maximize the impact of their investments.[9] (See Notes for details.) Many others have been working to develop evidence-based programs and policies.[10]

The rough blueprints for a parent nation have been drawn up. Now we need to build it. As I imagine it, one end of the early education and care continuum includes a cluster of policies that affect families during pregnancy and childbirth. Group prenatal care, for example, would bring expectant mothers together for care, providing community and more time for the anticipatory guidance that is so hard to get in our current system. There would be ample postnatal care as well, just as Katherine needed when she suffered postpartum depression. There would be much more widespread use of screening and referral programs for new parents, like Family Connects, which is already showing positive impact in places like Guilford County.

And then there's paid family and medical leave, one of the best and most effective ways to help families and children: It improves educational outcomes in the long term; it engages both parents (where there are two parents) in the development of the child; and, in a neat double whammy, it addresses some of the needs of early childcare. Often seen as a perk of employment, paid leave has mainly been available to those who work for the nation's largest employers. But even among the highest earners (who are most likely to get paid leave as a professional benefit), only one-fifth have access to paid leave through an employer.[11] The only way to have paid leave for everyone no matter where they work—to get it for people like Kimberly Montez and Sabrina—is for policy makers to set up and mandate a social insurance program.

In California, the first state to mandate paid leave, the state government created a state disability insurance plan funded by employee contributions. Since it took effect in 2004, worker retention has improved. New mothers who take paid leave are more likely to return both to the workforce and to the same job. Workers who take leave when they need it—not just for the birth of a child but to care for an ailing relative—are less stressed, and that improves productivity. Nearly 90 percent of California companies reported no increased costs because of the program, and 9 percent saved money because the paid leave program reduced turnover.[12]

The next cluster of policies in the continuum might seem to have nothing to do with early childhood development, but groups like

Cynthia's put them high on their lists because of how effective they have proven to be in improving families' overall well-being. They include tax credits for families, a living wage, and broader access to health insurance and SNAP (nutrition supplements). These are policies that help parents build sturdy boats to ferry their children through to adulthood and help them navigate the torrent when the wind whips up. Jade could have been the stay-at-home mother she dreamed of if her husband's job had provided health insurance. Randy needed wages that were sufficient to allow him to work one job instead of three.

Yet another cluster of programs is needed to support families in the years between birth and kindergarten. Chief among the areas to address is improving our dismal childcare system. That's what Gabby and Talia needed: quality, accessible care that they wouldn't have to take on extra jobs to afford. In fact, we need better options all around. We need to make it easier for those, like Jade, who want to stay home. And, for the large majority who are in the workforce, we need to provide childcare that truly serves the needs of young children and parents and is centered on nurturing interaction. Remember that, when the National Institute of Child Health and Human Development evaluated early childhood centers, it found that only 10 percent of them were delivering high-quality care.[13] But in looking for models of how to implement high-quality childcare at scale, I discovered that there is one American organization that is doing childcare extremely well: the military. At roughly eight hundred centers across the nation and overseas, the Department of Defense provides high-quality, universally accessible, and affordable child development centers serving children from birth to age twelve.[14] I would not have expected that innovative thinking in childcare would factor into thinking about national security, but it turns out that's exactly *why* it's being done.

About thirty years ago, military childcare was as bad as the worst we see today. "Stables, Quonset huts, asbestos-filled pre-WWII buildings with lead-based paint" is how M. A. Lucas, the woman who was brought in to change it, described the system she found in the early 1980s. "Care was custodial at best."[15] Many centers failed to meet safety standards; regulations and teacher training were paltry; and staff were paid less than anyone else on military bases, including the people who collected the trash. Turnover was as high as 300 percent. (It's now closer to 30. Still too high but a vast improvement, and lower than the turnover in just about any other childcare context.) In the late 1970s, my friend Linda Smith was charged with running a childcare center (they called it a "nursery") on an Air Force base in Arizona. Like M. A. Lucas, she was stunned at what she found when she arrived. Children were crowded into one room and babies were crowded together in playpens. There were no toys, just a television, and the children were all under the care (using the term loosely) of one untrained adult. "Obviously the challenges were enormous," Linda told me.

In a moment of clarity, military leaders realized that childcare was not just a problem for young children and their parents, this was a problem of military readiness. In short, the military had to figure out how to take care of its own, including record numbers of enlisted women, in a way that allowed them to excel at their jobs. As Lucas wrote years later, "It became clear that childcare or a lack of childcare could impact the nation's security and the military's ability to be ready to defend the country."[16]

The turnaround required an act of Congress. In the Military Child Care Act of 1989, Congress planned and funded a complete overhaul of the system. The changes began with large increases in teacher pay and training. Why start there? Because military leaders were galvanized by a 1989 study that showed higher wages were the number one predictor of quality in childcare.[17] Within twenty-four hours of the study's publication, Department of Defense officials were apparently on the phone with its authors, wanting to understand the relationship between pay and quality.[18] Just as Henry Ford found in his automobile factories, higher wages in childcare affect both who wants the job and who stays with it. In addition to raising pay, military leaders required stronger credentials for head teachers, stipulated training and curriculums based on the latest developmental thinking, and organized unannounced inspections every year to assess the overall quality and staff-child interactions. They also broke the link between what families can afford to pay and what it costs to provide high-quality services: Parents pay what they can afford, on a sliding scale—no one pays more than 10 percent of their income—and the military pays the difference. This is the opposite of how childcare works elsewhere in the United States, where most of the cost is passed on to families, and when they can't afford the fees, the quality of the childcare then available to them suffers.

"The most important lesson from the military is this—there is no single solution to a complex problem," Linda told me. After improving that childcare center in Arizona, she went on to replicate it on other bases and then worked with Lucas to overhaul the entire system. (Today, she directs early childhood policy for the Bipartisan Policy Center in Washington, DC.) In her view, the military's success came from seeing *all* the problems—professional development and pay, standards and their enforcement, the cost to families, and the availability of care—and addressing them head-on. "Many think that the funding solved all the problems—it was critical—but alone it would not have solved the problems," she said.

When military childcare bumps up against private care—for instance, when a military family must temporarily turn to a private childcare provider while a child is on the waiting list for a center on base—the differences become glaringly obvious. A technical sergeant who is a linguist in the Air Force sent all five of her children to military child development centers. "The [centers] are pretty much like a second home for my children," she told a reporter in 2017. "The teachers there are really invested in the kids." It was a real shock, then, when she had to put one of her kids in an off-base, private-sector center for a few months. It cost *weekly* what she spent over *several months* at the military center. "It was that big of a difference," the tech sergeant said. In talking with civilian colleagues, she realized how much the costs of childcare factor into most people's family decisions. "We just wanted to have kids," she said. "I felt like if I were a civilian that definitely would have had a huge impact on whether we kept having kids or not."[19]

In a 2013 report on childcare systems by Child Care Aware, the military got the highest grade awarded, a B. Twenty-one states got a D and some failed outright. "It's good news that the children of military families are in quality childcare," the report said. "It would be great news if the rest of the children in America could also be in quality care."[20]

Notably, the military system extends all the way from birth through pre-K and then into afterschool care for children until they hit thirteen. It neatly provides the continuum (including universal pre-K) that links early childhood to our K–12 system. Strong early childhood education and care lead to children who are ready to learn when they arrive in school.

The military example isn't perfectly replicable for the rest of society. But the quality, resources, and commitment that the military brought to early childhood education are exactly what's needed. And the fact that such a turnaround occurred within a major governmental institution is encouraging. "It's not a miracle," said one expert of the military's dramatic change. "It's the determination. And you have to fund it."

There is no way around it—you can't create quality programs without money. Most of what I have described will require increased investment at every level of government. We have fallen short repeatedly on this front. Our national spending on early childhood, by which I mean programs and services that affect children and their families, is around 0.2 percent of our gross domestic product, compared to an average of 0.7 percent in our peer nations.[21] That spending is primarily for subsidized childcare. The U.S. spends an average of $500 per child per year while other rich nations spend an average of $14,000 per child per year.[22] (This is just part of how other nations spend so much more on social services than we do.) At the state level, funding for early childhood has sometimes been linked to unrelated funding streams or "sin taxes"—like tobacco taxes in California, which then reduced revenue as the numbers of smokers dropped.[23] (While it's great news that smoking decreased, our progress on one public health problem caused us to lose ground on another.)

Many states are currently using the extra funds they received as pandemic relief to bolster their supports for children and families, but much of that funding is temporary. We cannot build a parent nation without stable funding solutions. Furthermore, investment in early childhood will pay dividends in the short term by getting more parents into the workforce and in the long term by building strong and productive future generations. It ultimately *saves* us money. Of course, Linda is certainly correct that there is no single solution and funding alone wouldn't have solved the problems the military was facing. In addition to increasing our spending, we must choose carefully what to invest in and then manage our programs wisely. But without funds, we will never move forward.

A Seat at the Table

Jovanna Archuleta knows firsthand how hard it is to survive as a young mother when you don't have any money, and how hard our current system can be to navigate. She was nineteen and living in Albuquerque when she got pregnant with her son, Ayden. She was about to begin college, the first in her family to do so, and she had cobbled together scholarships and loans to pay for school. Having a baby on her own was not part of the plan. Without her mother, Jovanna is not sure she could have done it. "My mom was there for me. She provided reassurance, communication, and commitment."

To have the baby, Jovanna went home to Nambé Pueblo, fifteen miles north of Santa Fe. When Ayden was six months old, she took him with her back to Albuquerque to begin school. By any measure, she was poor. Food stamps kept them fed, and she relied on Early Head Start for childcare. "We were in survival mode." Yet she remembers much about those early years with her son fondly. "It was probably the best time of our lives," she said. When she wasn't in class she was with Ayden. "We did the things that didn't cost money," she said. "We would go to the parks a lot. We would make tents in the living room and watch movies. Those were the little things that built our relationship." As she recalled those days for me, sixteen years later, her eyes welled with tears. "I get a little emotional," she said. "We say that our state and our people are in poverty, but that doesn't mean they're miserable. It doesn't stop you from giving those love experiences to your child." What did make her miserable was the way she was treated by the agencies whose services she so desperately needed. "You go into these systems being judged, you're no good. It was a horrible experience and a very time-consuming experience," she said. "Every time, it almost made me want to give up on getting through school."

Jovanna didn't give up. When she graduated with her BA, Ayden "graduated" from Head Start. Then she went on to get an MBA, got married, and had another child, Lily, who was three when we spoke (Ayden was sixteen). It was while she was pregnant with Lily that she began to work in early childhood ("I walked the walk") through a job with a private foundation, supporting early childhood efforts in New Mexico's Indigenous communities. In 2020, she was recruited for her new job as the first-in-the-nation statewide assistant secretary for Native American early childhood education and care, representing the people of the twenty-three New Mexican tribes (nineteen Pueblos, three Apache tribes, and the Navajo Nation).[24] Today, as assistant secretary, Jovanna is not just raising her own children, she has a hand in raising all the Native American children in New Mexico.

Her new boss, Elizabeth Groginsky, is also the first in *her* job. A veteran of the early childhood policy world who started in Head Start and strengthened universal pre-K in Washington, DC, by improving quality and increasing access, she was named New Mexico's first cabinet secretary for the Early Childhood Education and Care Department when it launched. She has her work cut out for her. On nearly every measure of early childhood health and education (child poverty, child well-being, and so on), the state ranks close to the bottom.[25] The secretary was brought in to change that after New Mexico became only the fourth state to create a cabinet-level agency dedicated to early childhood. (We don't have one in the federal government—yet!)

The new department, the secretary told me, "is a game-changer." Traditionally, early childhood has been a second-tier issue, or more precisely, it's been a small piece of other cabinet-level departments' responsibilities—child welfare, education, health, etc. Moreover, programs for early childhood exist in pieces across local, state, and federal governments, resulting in both gaps and overlapping jurisdiction. Having a dedicated department streamlines and simplifies the infrastructure. And it finally gives infants, toddlers, and their parents a metaphorical seat at the table where decisions about priorities and funding are made. "You've got to have an access point, to reach high-level government officials about issues affecting families in this birth-to-5 period," the secretary says. "For too long, the voice of early childhood [has not been] at any of those big decision-making tables."

I imagine that when tribal leaders requested that New Mexico's new early childhood department include someone dedicated to their communities, it was because they wanted one of their own to have a seat at the table, too. That person would understand the centrality of Native spirituality and cultural beliefs. And that person would know—intimately—the country's disgraceful history of removing Indian children from their families and communities, and the trauma and mistrust it sowed. Jovanna was their handpicked candidate.

As assistant secretary, with sensitivity and science on her side, Jovanna is using her new position to strengthen supports for families in Native communities in ways that also strengthen what is unique to Native culture. "Tribal communities [believe] each baby is given to their community as a gift to help them preserve what they need to be," Jovanna explained to me. "It's the children that are going to preserve Native people going forward." Here, too, there is a chance to get it right, finally. Far from a one-size-fits-all approach, Jovanna has been sharing the importance of foundational brain development with tribal leaders and looking for guidance from them on ways to make sure early childhood programs honor Native culture, aspirations, and spiritual beliefs. Among the projects she supports are language immersion preschools where very young children are learning Native languages. Those schools are thriving. Engaging, brain-enriching language experiences can come in any language so long as they are delivered by fluent speakers.

Her own Nambé Pueblo has directly felt the way early experiences affect health and well-being later in life. In collaboration with the Pueblo's governor, Phillip Perez, she is working to bridge the gap there in early childhood services. The Pueblo recently completed construction on a new child development center that will help Nambé's children get off to a strong start and by extension make the entire society healthier. "If we can build that foundation with our babies, and our families, and secure those attachments, those nurturing relationships," Jovanna said, "I think we are on a path to healing."

New Mexico's culturally specific approach for its tribal communities is a potent example of the way early childhood programs can be adapted to serve *all* communities—provided those programs are grounded in nurturing, rich serve-and-return interactions. Jovanna's spirituality came to mind when I spoke to Rachel Anderson, who leads a Christian group with roots in the evangelical community. Called Families Valued, the group is working to enact policies and programs like paid leave, child tax credits, and childcare.[26] Rachel, too, would like to be sure that parents in her community have choices that reflect their beliefs. That will mean working with faith-based organizations as providers of some services. But mostly, it means recognizing, as Rachel puts it, "the centrality of family life to human flourishing and the dignity of all forms of work." She wants to see support for those who choose to stay home and for those who work outside the home. The role of government, she says, is not to "handle functions of the family nor to control families' cultural and religious decisions."[27] But what government can do, in her view, is establish broad standards that push us toward pro-family workplace models and that support maternal and child health.

Like so many other people, Rachel was called to work on these issues by her own experience of becoming a parent and trying to get paid leave and health insurance. "It just shocked me that in a nation that values families so much, there was so little provision for starting a family and so little was anticipated about becoming a parent," Rachel said. "That reality felt deeply alienated from what I understand from my tradition, that families are a place that we honor the sacredness of life in all its forms. In many ways, families should shape the other structures in our society, or at least have equal claim on them." That is the heart of the matter.

In Rachel's work and in her words, I see another national

ideal—that we are a country made up of different faiths, different races, different ethnicities, different family configurations, and that our differences can make us stronger together. Our parent nation will be the same. By creating true choice for families, we honor our diversity. As Rachel said, "The public square is comprised of people and institutions whose values and beliefs are different. And yet, we're all called to work together in this democracy to steward the common good."

The Urgency of Now

In the two years that have passed since Michael was released, he hasn't always found it easy to readjust to life outside jail. There have been challenges large and small. The first time he and Keyonna went shopping on a busy street in Chicago, Michael froze trying to cross the street. "The cars are moving so fast," he told Keyonna, who had dashed out into the street like she always did. She came back and took his hand and led him across.

Hand in hand, they have persevered. "I feel great being back around my family, back in my comfort zone," Michael told me over Sunday brunch in the summer of 2021. Desperate to make up for lost time, he recalled how he had spent the first weeks of his freedom doing everything he could think of with his son: visiting the zoo, going on a boat trip, playing the video games that had come out while he was away (Mikeyon loved being the expert).

Michael quickly got a job at another meat wholesaler and has already been promoted to assistant to the manager. Keyonna now works at the Chicago Housing Authority. She, too, has been promoted and won plaudits for her work. Through a housing voucher, they were eventually able to move to Wicker Park, a safer neighborhood than North Lawndale, where their kids can walk to school without fear.

But the happiest part of their story is Mikeyon. In kindergarten, he was reading at a second-grade level. For first grade, they were figuring out, when we met, if he would go to the neighborhood school or a magnet school. Clearly, Keyonna's loving interactions with her son gave him what he needed. Wonderful though Mikeyon's progress is, I can't help but think how hard it was for Keyonna to do it alone. Michael lost his chance to be a brain architect during the critical first five years of Mikeyon's life.

When I talked with Michael and Keyonna over that brunch, I could hear the urgency they feel now that they are parenting together. It is the urgency that comes with realizing how precious time is, how much time with our children matters. They are bursting with pride over Mikeyon. They love to listen to their son's ideas about what he will be when he grows up—one day it's a basketball star, the next an architect. (I'm rooting for doctor.) Despite everything they've been through, they are hopeful about the future.

As I write, Congress is considering major new legislation that would help families. According to Cynthia, ten states have passed legislation to provide paid leave, and twenty others introduced such legislation during the pandemic. It didn't pass everywhere, but there is more momentum at the state level to support families in ways we have not been in the past. "The pandemic created just immeasurable crises for so many folks," Cynthia says. And government rose to the challenge of helping those people. "It demonstrates we can do this. If we lay the foundation and show that it is having an impact, my hope is that we'll make a lot of what we have implemented on a temporary basis permanent."

We must lay those foundations in society and in the brains of our children. On that front, I share both Michael and Keyonna's hope and their urgency.

Fifty years ago, the Comprehensive Child Development Act (CCDA) came within a pen stroke of being signed into law. We stood at a crossroads, and we turned back.

Today, we are at a new crossroads. This time we must forge ahead.

To borrow one more time from the stirring words of Martin Luther King Jr., we are once again "confronted with the fierce urgency of now."

And it is truly urgent. Because while not enough has changed in fifty years, there are some things that have.

What we now know of the brain demands urgency. In 1971, when the CCDA was being formulated, people were just beginning to understand how critical the early years of experience are to children's development. Since then, we have learned so much more. There is neuroscientific heft about what children need, when they need it, and the essential role of parents and caregivers as children's first, best teachers.

What we experienced during the pandemic demands urgency. While the novel coronavirus presented some unavoidable challenges, it revealed the inescapable fallacy of the idea that anyone can successfully parent alone. And it showed us that our professional and personal lives are intertwined. We can no longer pretend they aren't.

Finally, our national ideals demand urgency. It is time to live up to our claims of opportunity and equality for all. Let's not accept the fading of our American Dream. Let's reclaim it and let's build a parent nation.

EPILOGUE

"Of all the stars in this 'brave, overhanging sky,' the
North Star is our choice . . .

"To millions now in our boasted land of liberty, it
is the star of hope."

—FREDERICK DOUGLASS[1]

As the years passed after Don's death, the raging torrent began to settle and the rising sun slowly forced the darkness from the sky. The river's edge of young adulthood came slowly into view. My children's three small, terrified faces no longer peered up at me. Instead, the mature and joyful faces of Genevieve, Asher, and Amelie gazed down at me. They carried their father's memory and empathy—and his height—into young adulthood.

I continued forward, alone, retracing the steps each morning that Don and I had taken to work. My research, like my children, remained a source of solace, purpose, and continued learning. As a self-trained social scientist (though a formally trained surgeon), I learned from everyone around me. I learned from the families I worked with and cared for. I learned from my colleagues across the university, country, and world. Understanding the complexities of what needed to happen in order to safeguard the promise of every child's promise was my life's work.

As fate would have it, I met a brilliant economist, John List, for whom the issues of children held a special place.

Economists bring to life the humorous University of Chicago saying, "That's all well and good in practice . . . but how does it work in theory?" They create simplified equations or "models," which strip away the "noise," to reveal life's fundamental truths. As we began to work together, he showed me that the answer I sought was, in fact, quite simple.

It wasn't all theory for this economist. A devoted father who'd never known a moment of "secret parenting," he was unabashedly raising *five* wonderful children, Annika, Eli, Noah, Greta, and Mason (in tandem with his wonderful ex-wife, Jen). He coached his kids' baseball teams (Cooperstown champions!). He enticed them to watch *The Sound of Music*, his personal favorite, repeatedly. He introduced them to the "joys" of economics, even if some of them considered that an oxymoron. And he inspired them to think about what they wanted to contribute to the world. His goal was simple: to make memories filled with love.

Over time, these memories became *our* memories, as first our work, then our lives and our families began to intertwine.

As John's love enveloped Amelie, Asher, and Genevieve, I imagined, with mixed emotions, Don watching over his beloved children. John's presence was in some ways a painful reminder that while Don would never be replaced, he did not live to see his children grow up. But I was consoled by the peace and joy I knew Don would feel in knowing that his children were loved and cared for, and that I had help navigating the torrent. John was honoring Don in the most profound of ways. He loved Don's children as *Don* would have wanted. He cherished Don's ultimate legacy. My children and his children have become our children.

With profound happiness, John and I formalized our union. Our nuptials were splashed across the front page of the University of

Chicago *Maroon*, under a headline that trumpeted: "From Thirty Million Words to Just Two: I Do!" As we exchanged vows on that beautiful fall day, I realized that this economist had indeed led me to the answer I'd been seeking. It was a feeling that I had been seeing all along in the faces of my patients' parents when I took their babies into my arms and carried them into surgery. He had created the perfect "model" for understanding what is fundamentally necessary to allow all children to reach the promise of their promise.

It is the ability to see someone else's child as your own, and to support their parents as they strive to cross the raging torrent.

My scientific journey, too, has taken me from thirty million words to just two: our children.

• BUILDING A PARENT NATION •
ACTION GUIDE
Foster community. Forge collective identity.
Fight for change.

Now that you've finished *Parent Nation*, you may feel called to help build one. Good news! This guide can help.

The first and most important step in building a nation that works with and for parents is a mindset shift. Parents are a beautifully diverse group. We each see the world through our own lens, and we bring our lived experiences to the sacred task of raising children. But we're also a collective linked by a fierce love for the humans we're raising and a boundless hope for their futures. That collective identity can inspire us to grant all parents empathy and a sense of community and belonging. I may not look or live like you do, but I love like you do. And that makes us partners.

This collective identity gives us strength, and we can use that strength to ensure we receive the systemic support we all need. Parents cannot and should not be expected to go it alone when engaged in the important work of raising children. We can, and should, invite our workplaces, communities, and country to provide parents the necessary supports so that children are nurtured from their earliest days of life. The science demands as much. Our children deserve as much.

If you feel called to create change on behalf of parents and children, you may consider one of the following paths.

- Visit ParentNation.org to learn about organizations around the country working on behalf of parents and children. If you're able, support one with your time or donation.
- Read and sign on to the Parent Nation Ideals.
- Share your story on ParentNation.org.
- Use your social media platforms to share the work of organizations working on behalf of parents and children and invite others to do the same.
- Create a parent advisory board in your community spaces (your workplace, your house of worship, your school district) to ensure all parents' voices and needs are being equally heard.
- Talk to parents in your neighborhood, town, or city about their experiences and bring their concerns and ideas to your local elected officials.
- Conduct a survey of parents in your workplace to find out what supports would most benefit them, and share the results with leadership.
- Write a letter to the editor of a local or national newspaper advocating for a project or policy that supports parents and children.
- When elections come around, research policies and candidates with an eye toward what's best for parents and children. Vote!
- Form a book club to discuss *Parent Nation*, or suggest the book to your existing group. (See the discussion guide section.)
- Create or join a Parent Village. Parent Villages are small groups of parents who come together to support one another, identify and discuss the needs of families in their community, and make a plan to get those needs

met. Your Village may be a group of coworkers, a collection of parents from your child's school, a group of your neighbors, or fellow parishioners at your house of worship. Any combination of up to ten committed change agents will do. You can find everything you need, including a suggested curriculum for your Village meetings, at ParentNation.org/Villages.

• DISCUSSION GUIDE •

Note to readers: If you're using this guide for yourself or a small group of two or three people, you may want to tackle each discussion point. If you're discussing the book with a larger group, feel free to pick one or two discussion points from each chapter.

ONE: Toward a New North Star

Discussion points:

- Do you have the tools and teammates you need to raise your child(ren)?
- If not, what are the barriers to finding or keeping those tools and teammates?
- What or who could help you remove those barriers?
- Are you able to help remove them for other parents in your community?
- What were your biggest parenting challenges during the pandemic?
- Have you ever felt ashamed to acknowledge that parenting feels overwhelming?
- Do you think of other parents as your allies?

TWO: The Brain's Greatest Trick

Discussion points:

- What did Charlotte's story teach you about the brain's ability to rewire itself?

- How might her story influence the way we approach childcare and early childhood education?
- What did Hazim's story teach you about the distribution of resources and opportunity in the United States?
- Did society help or hamper Charlotte's road to success? How about Hazim's?
- Imagine a society that viewed foundational brain development as its North Star. What would have to change in our schools? Our workplaces? Our communities?
- Are you able to help advocate for any of those changes?

THREE: The Streetlight Effect

Discussion points:

- What options did your parents have for your care before you started full-time school?
- How did those options differ from what was available when you became a parent?
- What are some ways that society could increase wages for childcare workers without simply shifting that cost to parents?
- If you could reimagine our school system, based on twenty-first-century thinking, what's the first thing you would change?
- Some parents can and want to provide their children's full-time care until they begin school. What supports do those parents need?
- Some parents need to or prefer to work full-time outside the home. What supports do those parents need?

FOUR: The Brain Architects

Discussion points:

- Spend a few moments considering or discussing with your group the "veil of ignorance" thought experiment.
- In terms of parenting, how would your imagined society differ from the current one in the United States?
- How would it be similar?
- Do you know parents (including yourself) whose dreams for how they'd parent were sidelined by unexpected barriers?
- Kimberly took her story to the American Academy of Pediatrics, where she helped craft a policy statement on the importance of paid family medical leave. What could you do with the parent stories you've heard or lived?

FIVE: It All Starts with Beliefs

Discussion points:

- When and where did you first learn about what babies need to develop healthy brains?
- Did it change how you interacted with your child? If this information is new to you, will it change how you interact with your child going forward?
- What are some ways we could pass this information along to parents earlier and more systemically?
- Based on the author's descriptions, do you identify as a "concerted cultivation" parent or a "natural growth" parent?
- Suskind writes that the disconnect between what we say and what we do as a nation has direct—and sometimes dire—consequences for parents. Can you think of an example? Can you think of a solution?

SIX: Building Foundations and Building Sturdy Boats

Discussion points:

- Have there been moments when you wished you had more time to engage with your child?
- In what ways do economic pressures affect your parenting?
- How much did the cost of childcare impact your decisions about when and whether to return to work after becoming a parent?
- Were you able to find affordable childcare that also offered the attention you wanted for your child?
- What are some other ways workplaces could work with, rather than against, parents?

SEVEN: Making Maps and Navigating the Torrent

Discussion points:

- What similarities do you see between Sabrina's and Katherine's stories?
- Do you recognize yourself in either of the women or their families? Do you recognize a parent you know?
- Is your community set up to help parents provide a calm, stable environment for their children? What about the communities surrounding yours?
- Katherine envisions intentional parenting communities, where parents from all different backgrounds come together and support one another. What would that look like where you live?
- Research shows that countries with fewer robust family support policies are more likely to show disparities in the health of children. What policies should be put in

place so that parents receive the support Katherine dreams of, but on a systemic level?

EIGHT: Lifting Our Voices

Discussion points:

- What would a modern version of the Comprehensive Child Development Act include in order to support American families?
- What are some lessons we could draw from AARP's success in order to improve the health and well-being of parents and children?
- Should healthy brain development be considered a civil right? Why or why not?
- The author writes, "Sometimes we are limited by where we set our sights. Then, suddenly, we see over the horizon and understand there is another way." Does that other way feel reachable to you? Why or why not?

NINE: Just What the Doctor Ordered

Discussion points:

- Have your healthcare providers (ob-gyn, pediatrician) been reliable sources of information about what your child's brain needs?
- Where would you typically turn to find resources or advice about building your child's brain?
- Are such resources easy to access?
- How does the United States' ratio of dollars spent on medical care versus social services strike you? Should we spend more on social services? Less? About what we currently do? Why?

TEN: The Business of Business Is . . .

Discussion points:

- Have you ever felt the need to hide your family's needs from your employer?
- How would it be received if you told your employer you needed to leave by a certain time to see your child's baseball game/recital/school play or take your child to the doctor?
- Have you engaged in conversations with coworkers about whether your workplace is conducive to parenting? If so, have you shared any of those conversations with higher-ups?
- Do you think workplaces can balance both profits and their employees' personal lives?
- Political scientist Jacob Hacker said, "The family used to be a refuge from risk. Today, it is the epicenter of risk." What changes brought about that shift?

ELEVEN: Life, Liberty, and Fulfilling Society's Promise

Discussion points:

- The Military Child Care Act of 1989 reimagined and reshaped childcare for families in the service. Can you imagine something similar working in the civilian world?
- Do issues such as paid leave, child tax credits, and childcare strike you as inherently political? Why or why not?
- The author introduced us to many parents throughout the book. Whose story resonated most with you and why?

- Do you feel a sense of belonging or collective identity with other parents in a way you didn't before reading the book?
- Did the book leave you feeling hopeful? Why or why not?

• NOTES •

CHAPTER ONE

1. Mandela, N. (1995). Address by President Nelson Mandela at the launch of the Nelson Mandela Children's Fund, Pretoria.
2. Hart, B., & Risley, T. R. (1995). *Meaningful differences in the everyday experience of young American children.* Baltimore: Paul H. Brookes Publishing.
3. Denworth, L. (2014). *I can hear you whisper: An intimate journey through the science of sound and language* (ch. 22). New York: Dutton; Romeo, R. R., Segaran, J., Leonard, J. A., Robinson, S. T., West, M. R., Mackey, A. P., . . . & Gabrieli, J. D. (2018). Language exposure relates to structural neural connectivity in childhood. *Journal of Neuroscience, 38*(36), 7870–77.
4. Hoff, E. (2013). Interpreting the early language trajectories of children from low-SES and language minority homes: Implications for closing achievement gaps. *Developmental Psychology, 49*(1), 4; Dickinson, D. K., & Porche, M. V. (2011). Relation between language experiences in preschool classrooms and children's kindergarten and fourth-grade language and reading abilities. *Child Development, 82*(3), 870–86.
5. Hirsh-Pasek, K., Adamson, L. B., Bakeman, R., Owen, M. T., Golinkoff, R. M., Pace, A., . . . & Suma, K. (2015). The contribution of early communication quality to low-income children's language success. *Psychological Science, 26*, 1071–83.
6. History of TMW Center. https://tmwcenter.uchicago.edu/tmwcenter/who-we-are /history.
7. Suskind, D. L. (2015). *Thirty million words: Building a child's brain* (pp. 246–47). New York: Dutton.
8. Suskind, D. L., Leung, C. Y., Webber, R. J., Hundertmark, A. C., Leffel, K. R., Fuenmayor Rivas, I. E., & Grobman, W. A. (2018). Educating parents about infant language development: A randomized controlled trial. *Clinical Pediatrics, 57*(8), 945–53; Leung, C. Y., Hernandez, M. W., & Suskind, D. L. (2020). Enriching home language environment among families from low-SES backgrounds: A randomized controlled trial of a home visiting curriculum. *Early Childhood Research Quarterly, 50*, 24–35.
9. United States Census Bureau. (2021, October 8). America's Family and Living Arrangements: 2020. https://www.census.gov/data/tables/2020/demo/families /cps-2020.html.
10. OECD Family Database, Parental Leave Systems. https://www.oecd.org/els/soc /PF2_1_Parental_leave_systems.pdf.

11. Malik, R., Hamm, K., Schochet, L., Novoa, C., Workman, S., & Jessen-Howard, S. (2018). America's child care deserts in 2018. Report of Center for American Progress, pp. 3–4; National Institute of Child Health and Human Development (2006). The NICHD study of early child care and youth development: Findings for children up to 4½ years.

12. Bureau of Labor Statistics, U.S. Department of Labor. Employment characteristics of families—2020. Press release, Apr 21, 2021.

13. Donovan, S. A., & Bradley, D. H. (2018, Mar 15). Real wage trends, 1979 to 2017. Washington, DC: Congressional Research Service [cited 2019, Apr 9].

14. Schleicher, A. (2019). *PISA 2018: Insights and interpretations*. Paris: OECD Publishing.

15. Chan, J. Y., Wong, E. W., & Lam, W. (2020). Practical aspects of otolaryngologic clinical services during the 2019 novel coronavirus epidemic: An experience in Hong Kong. *JAMA Otolaryngology–Head & Neck Surgery, 146*(6), 519–20. doi:10.1001/jamaoto.2020.0488.

16. Kalil, A., Mayer, S., & Shah, R. (2020, Oct 5). *Impact of the COVID-19 crisis on family dynamics in economically vulnerable households*. University of Chicago, Becker Friedman Institute for Economics, Working Paper No. 2020-143, available at SSRN: https://ssrn.com/abstract=3706339 or http://dx.doi.org/10.2139/ssrn.3706339.

17. Lee, E. K., & Parolin, Z. (2021). The care burden during COVID-19: A national database of child care closures in the United States. *Socius*. https://doi.org/10.1177/23780231211032028.

18. Shrimali, B. P. (2020). *Child care, COVID-19, and our economic future*. Federal Reserve Bank of San Francisco Community Development Research Brief 2020-5.

19. Heggeness, M., Fields, J., García Trejo, Y. A., & Schulzetenber, A. (2021). Tracking job losses for mothers of school-age children during a health crisis. United States Census Bureau.

20. García, J. L., Heckman, J. J., Leaf, D. E., & Prados, M. J. (2020). Quantifying the life-cycle benefits of an influential early-childhood program. *Journal of Political Economy, 128*(7), 2502–41; Executive Office of the President of the United States (2014, Dec). *The economics of early childhood investments*.

21. ReadyNation Report (2019, Jan). Want to grow the economy? Fix the child care crisis; S&P Global (2020, Oct 13). Women at work: The key to global growth.

22. UNICEF. (2019). Family-friendly policies: Redesigning the workplace of the future: A policy brief.

23. King, M. L., Jr. "I've been to the mountaintop" speech, Apr 3, 1968. Memphis, TN.

24. Kisilevsky, B. S., Hains, S. M., Lee, K., Xie, X., Huang, H., Ye, H. H., . . . & Wang, Z. (2003). Effects of experience on fetal voice recognition. *Psychological Science, 14*(3), 220–24.

25. Kolb, B. (2009). Brain and behavioural plasticity in the developing brain: Neuroscience and public policy. *Paediatrics & Child Health, 14*(10), 651–52; Fox, S. E., Levitt, P., & Nelson, C. A. III (2010). How the timing and quality of early experiences influence the development of brain architecture. *Child Development, 81*(1), 28–40.

26. Hoff, E. (2013). Interpreting the early language trajectories of children from low-SES and language minority homes: Implications for closing achievement

gaps. *Developmental Psychology, 49*(1), 4; Weisleder, A., & Fernald, A. (2013). Talking to children matters: Early language experience strengthens processing and builds vocabulary. *Psychological Science, 24*(11), 2143–52.

27. Nelson, C. A., Zeanah, C. H., Fox, N. A., Marshall, P. J., Smyke, A. T., & Guthrie, D. (2007). Cognitive recovery in socially deprived young children: The Bucharest Early Intervention Project. *Science, 318*(5858), 1937–1940; Lund, J. I., Toombs, E., Radford, A., Boles, K., & Mushquash, C. (2020). Adverse childhood experiences and executive function difficulties in children: A systematic review. *Child Abuse & Neglect, 106*, 104485.

CHAPTER TWO

1. Tate, C. (Ed.) (1985). *Black women writers at work* (p. 7). England: Oldcastle Books.
2. Asaridou, S. S., et al. (2020). Language development and brain reorganization in a child born without the left hemisphere. *Cortex, 127*, 290–312.
3. Isaac Asimov speaking to students at Trinity College, Hartford, CT, April 20, 1988.
4. Sakai, J. (2020). Core concept: How synaptic pruning shapes neural wiring during development and, possibly, in disease. *Proceedings of the National Academy of Sciences, 117*(28), 16096–99; Huttenlocher, P. R. (1979). Synaptic density in human frontal cortex—Developmental changes and effects of aging. *Brain Research, 163*(2), 195–205; Center on the Developing Child, Harvard University. Brain architecture. https://developingchild.harvard.edu/science/key-concepts/brain-architecture/.
5. Fox, S. E., Levitt, P., & Nelson, C. A. III (2010). How the timing and quality of early experiences influence the development of brain architecture. *Child Development, 81*(1), 28–40.
6. Niparko, J. K., Tobey, E. A., Thal, D. J., Eisenberg, L. S., Wang, N. Y., Quittner, A. L., . . . & CDaCI Investigative Team. (2010). Spoken language development in children following cochlear implantation. *JAMA, 303*(15), 1498–506.
7. Knickmeyer, R. C., Gouttard, S., Kang, C., Evans, D., Wilber, K., Smith, J. K., . . . & Gilmore, J. H. (2008). A structural MRI study of human brain development from birth to 2 years. *Journal of Neuroscience, 28*(47), 12176–82.
8. Shonkoff, J. P., Boyce, W. T., & McEwen, B. S. (2009). Neuroscience, molecular biology, and the childhood roots of health disparities: Building a new framework for health promotion and disease prevention. *JAMA, 301*(21), 2252–59; Shonkoff, J. P. (2012). Leveraging the biology of adversity to address the roots of disparities in health and development. *Proceedings of the National Academy of Sciences, 109* (Suppl 2), 17302–7; National Scientific Council on the Developing Child (2020). *Connecting the brain to the rest of the body: Early childhood development and lifelong health are deeply intertwined.* Working Paper No. 15.
9. Crime: García, J. L., Heckman, J. J., & Ziff, A. L. (2019). Early childhood education and crime. *Infant Mental Health Journal, 40*(1), 141–51. Income: García, J. L., Heckman, J. J., Leaf, D. E., & Prados, M. J. (2020). Quantifying the life-cycle benefits of an influential early-childhood program. *Journal of Political Economy, 128*(7), 2502–41.
10. Asaridou, S. S., et al. (2020).

11. Siman-Tov, M., Radomislensky, I., Knoller, N., Bahouth, H., Kessel, B., Klein, Y., . . . & Peleg, K. (2016). Incidence and injury characteristics of traumatic brain injury: Comparison between children, adults and seniors in Israel. *Brain Injury, 30*(1), 83–89.

12. Feuillet, L., Dufour, H., & Pelletier, J. (2007). Brain of a white-collar worker. *The Lancet, 370*(9583), 262.

13. Child Trends (2019, Jan 28). Children in poverty. https://www.childtrends.org /indicators/children-in-poverty.

14. Johnson, S. B., Riis, J. L., & Noble, K. G. (2016). State of the art review: Poverty and the developing brain. *Pediatrics, 137*(4); Chetty, R., et al. The association between income and life expectancy in the United States, 2001–2014. *JAMA, 315*(16), 1750–66. doi:10.1001/jama.2016.4226.

15. Noble, K. G., & Giebler, M. A. (2020). The neuroscience of socioeconomic inequality. *Current Opinion in Behavioral Sciences, 36*, 23–28; Luby, J. (2015). Poverty's most insidious damage: The developing brain. *JAMA Pediatrics, 169*(9), 810–11.

16. Noble, K. G., et al. (2015). Family income, parental education and brain structure in children and adolescents. *Nature Neuroscience, 18*(5), 773–78. See also Kimberly Noble's TED Talk from April 2019.

17. Schnack, H. G., et al. (2014). Changes in thickness and surface area of the human cortex and their relationship with intelligence. *Cerebral Cortex, 25*(6), 1608–17.

18. National Research Council (US) and Institute of Medicine (US) Committee on Integrating the Science of Early Childhood Development (2000). Ch. 8: The developing brain. In Shonkoff, J. P., Phillips, D. A. (Eds.), *From neurons to neighborhoods: The science of early childhood development*. Washington, DC: National Academies Press (US).

19. Suskind, D. L., Leffel, K. R., Graf, E., Hernandez, M. W., Gunderson, E. A., Sapolich, S. G., . . . & Levine, S. C. (2016). A parent-directed language intervention for children of low socioeconomic status: A randomized controlled pilot study. *Journal of Child Language, 43*(2), 366–406.

20. Noble, K. G., et al. (2012). Neural correlates of socioeconomic status in the developing human brain. *Developmental Science, 15*(4), 516–27; Noble, K. G., et al. (2021). Baby's first years: Design of a randomized controlled trial of poverty reduction in the United States. *Pediatrics, 148*(1).

21. Farah, M. J. (2018). Socioeconomic status and the brain: Prospects for neuroscience-informed policy. *Nature Reviews Neuroscience, 19*, 428–38.

22. Kim Noble TED Talk; Noble, K. G., Engelhardt, L. E., Brito, N. H., Mack, L. J., Nail, E. J., Angal, J., . . . & PASS Network. (2015). Socioeconomic disparities in neurocognitive development in the first two years of life. *Developmental Psychobiology, 57*(5), 535–51; Feinstein, L. (2003). Inequality in the early cognitive development of British children in the 1970 cohort. *Economica, 70*(277), 73–97.

23. In addition to Kim Noble's work, see Assari, S. (2020). Parental education, household income, and cortical surface area among 9–10 years old children: Minorities' diminished returns. *Brain Sciences, 10*(12), 956.

24. Troller-Renfree, S. V., et al. The impact of a poverty reduction intervention on infant brain activity. *PNAS*, in press.

25. Children's Defense Fund. (2020). *The state of America's children*. Annual report.

26. Attributed, among others, to Bill Clinton at the 2012 Clinton Global Initiative Forum.

27. Mooney, T. (2018, May 11). Why we say "opportunity gap" instead of "achievement gap." TeachforAmerica.org.

28. Shaughnessy, M. F., & Cordova, M. (2019). An interview with Jonathan Plucker: Reducing and eliminating excellence gaps. *North American Journal of Psychology, 21*(2), 349–59; Hardesty, J., Jenna McWilliams, J., & Plucker, J. A. (2014). Excellence gaps: What they are, why they are bad, and how smart contexts can address them . . . or make them worse. *High Ability Studies, 25*(1), 71–80.

29. Wyner, J. S., Bridgeland, J. M., & DiIulio, J. J., Jr. Achievement trap: How America is failing millions of high-achieving students from lower-income families. Jack Kent Cooke Foundation Report.

30. Rambo-Hernandez, K., Peters, S. J., & Plucker, J. A. (2019). Quantifying and exploring elementary school excellence gaps across schools and time. *Journal of Advanced Academics, 30*, 383–415; Plucker, J. (2013). Talent on the sidelines: Excellence gaps and America's persistent talent underclass. Storrs: Center for Education Policy Analysis, University of Connecticut. Retrieved from http://cepa.uconn.edu/mindthegap; Plucker, J. A., Burroughs, N. A., & Song, R. (2010). Mind the (other) gap: The growing excellence gap in K–12 education. Bloomington, IN: Center for Evaluation and Education Policy.

31. Moody, M. (2016). From under-diagnoses to over-representation: Black children, ADHD, and the school-to-prison pipeline. *Journal of African American Studies, 20*(2): 152–63; Gilliam, W. S. (2005). Prekindergarteners left behind: Expulsion rates in state prekindergarten systems. New York: Foundation for Child Development.

32. Quoted in Howard, J. (1963, May 24). Doom and glory of knowing who you are. *LIFE, 54*(21): 89.

33. Quoted in Snyder, S. (2018, Dec 19). North Philly to Oxford. *Philadelphia Inquirer.*

34. About Temple University. https://www.temple.edu/about.

35. Rhodes Trust announcement of Class of 2018, Dec 12, 2017.

36. Quoted in Snyder (2018, Dec 19). *Philadelphia Inquirer.*

37. UCL Centre for Longitudinal Studies.

38. Feinstein, L. (2003). Inequality in the early cognitive development of British children in the 1970 Cohort. *Economica, 70*(177), 73–97.

39. Van Dam, A. (2018, Oct 9). It's better to be born rich than gifted. *Washington Post.*

CHAPTER THREE

1. From the video game *Red Dead Redemption II.*

2. Battaglia, M., & Atkinson, M. A. (2015). The streetlight effect in type 1 diabetes. *Diabetes, 64*(4), 1081–90.

3. Begley, S. (2019, Jun 25). The maddening saga of how an Alzheimer's "cabal" thwarted progress toward a cure for decades. *STAT.* https://www.statnews.com /2019/06/25/alzheimers-cabal-thwarted-progress-toward-cure/.

4. John, L. K., et al. (2017, Mar–Apr). What's the value of a like? *Harvard Business Review.*

5. Freedman, D. H. (2010, Dec 9). Why scientific studies are so often wrong: The streetlight effect. *Discover.* https://www.discovermagazine.com/the -sciences/why-scientific-studies-are-so-often-wrong-the-streetlight-effect.

6. Lessons from high-performing countries: Secretary Duncan's remarks at National Center on Education and the Economy National Symposium, 2011.

7. Chetty, R., Grusky, D., Hell, M., Hendren, N., Manduca, R., & Narang, J. (2017). The fading American dream: Trends in absolute income mobility since 1940. *Science, 356*(6336), 398–406.

8. Committee of the Virginia Assembly. 79. A bill for the more general diffusion of knowledge, 18 June 1779. *Founders Online,* National Archives.

9. Schleicher, A. (2019). PISA 2018: Insights and interpretations. Paris: OECD Publishing.

10. Goldstein, D. (2019, Dec 3). It just isn't working: PISA test scores cast doubt on U.S. education efforts. *New York Times.*

11. Beatty, B. (1995). *Preschool education in America: The culture of young children from the colonial era to the present.* New Haven, CT: Yale University Press.

12. Comenius, J. A. (1907). *The great didactic.* trans. by M. W. Keatinge, Adam and Charles Black; Maviglia, D. (2016). The main principles of modern pedagogy in "Didactica Magna" of John Amos Comenius. *Creative Approaches to Research, 9*(1).

13. Hiatt, D. B. (1994). Schools: An historical perspective 1642–. *School Community Journal, 4*(2); Plucknett, T. F. (1930). The laws and liberties of Massachusetts. https://www.mass.gov/files/documents/2016/08/ob/deludersatan.pdf; Walker, B. D. (1984). The local property tax for public schools: Some historical perspectives. *Journal of Education Finance, 9*(3), 265–88.

14. Comenius, J. A. (1898). *Comenius's school of infancy: An essay on the education of youth during the first six years.* Ed. W. S. Monroe. (p. 81). Norwood, MA: Norwood Press.

15. White, S. H. (1996). The child's entry into the "Age of Reason." In A. J. Sameroff & M. M. Haith (Eds.) *The five to seven year shift: The age of reason and responsibility* (pp. 17–30). Chicago: University of Chicago Press.

16. Minkeman, P. (2014). Reforming Harvard: Cotton Mather on education at Cambridge. *The New England Quarterly, 57*(2); Sewall, S. (1969). *The selling of Joseph: A memorial.* Ed. S. Kaplan. Boston: University of Massachusetts Press; Sewall, S. (1882). *Diary of Samuel Sewall: 1674–1729.* Massachusetts Historical Society; Graham, J. S. (2000). *Puritan family life: The diary of Samuel Sewall.* Boston: Northeastern University Press.

17. Moran, G. F., & Vinovskis, M. A. (1985). The great care of godly parents: Early childhood in Puritan New England. In A. B. Smuts & J. W. Hagen (Eds.) *History and research in child development. Monographs of the Society for Research in Child Development, 50*(4–5, Serial No. 211), 24–37.

18. Mather, C. (1689). Small offers towards the service of the tabernacle in this wilderness (R. Pierce), pp. 59–61. Early English Books Text Creation Partnership Online, 2011.

19. Earle, A. M. (1899). *Child life in colonial days.* Darby, PA: Folcroft Library Editions.

20. Graham, J. S. (2000), pp. 111–14.

21. Suneson, G. (2019, Apr 4). What are the 25 lowest paying jobs in the US? Women usually hold them. *USA Today*. See also Occupational and Employment Wages. U.S. Bureau of Labor Statistics (2020, May). Childcare workers.

22. Mondale, S., & Patton, S. B. (2001). *School: The story of American public education* (p. 118). Boston: Beacon Press.

23. Hans, N. (2012). *Comparative education: A study of educational factors and traditions*. Routledge; Peters, V. (1956). "Education in the Soviet Union." *The Phi Delta Kappan, 37*(9), 421–25.

24. National Commission on Excellence in Education. (1983). A nation at risk: The imperative for educational reform. Washington, DC: U.S. Government Printing Office; Mehta, J. (2015). Escaping the shadow: A nation at risk and its far-reaching influence. *American Educator, 39*(2), 20; Mondale, S., & Patton, S. B. (2001), p. 177.

25. McCartney, K., & Phillips, D. (1988). Motherhood and child care. In B. Birns & D. Hay (Eds.). *The different faces of motherhood* (p. 160). New York, NY: Springer Science and Business Media.

26. Cohen, A. J. (1996). A brief history of federal financing for child care in the United States. *The Future of Children, 6*(2), 26–40.

27. Roosevelt, E. (2017). My Day, September 8, 1945.

28. Child care center opened by mayor. (1943, Jan 26). *New York Times*, p. 16.

29. All cited in Shonkoff, J. P., & Meisels, S. J. (1990). Early childhood intervention: The evolution of a concept. *Handbook of early childhood intervention*. (pp. 13–16). Cambridge University Press.

30. Thompson, O. (2018). Head Start's long-run impact: Evidence from the program's introduction. *Journal of Human Resources, 53*(4), 1100–1139. https://doi.org/10.3368/jhr.53.4.0216-7735R1.

31. Children's Defense Fund. (2020). *The state of America's children*. Annual report.

32. Wrigley, J. (1989). Do young children need intellectual stimulation? Experts' advice to parents, 1900–1985. *History of Education Quarterly, 29*(1) (Spring), 41–75.

33. Bureau of Labor Statistics (2002). A century of change: The U.S. labor force, 1950–2050. Monthly Labor Review, May; Rothwell, J., & Saad, L. (2021, Mar 8). How have U.S. working women fared during the pandemic? Gallup.

34. Organisation for Economic Co-operation and Development (OECD). (2020). *Early learning and child well-being: A study of five-year-olds in England, Estonia, and the United States*. Paris: OECD.

35. European Commission, EURYDICE, Finland. (2021, Apr). Early childhood education and care.

CHAPTER FOUR

1. Bowlby, J., & World Health Organization. (1952). Maternal care and mental health: A report prepared on behalf of the World Health Organization as a contribution to the United Nations programme for the welfare of homeless children (2nd ed., p. 84). World Health Organization.

2. Trevathan, W. R., & Rosenberg, K. R. (Eds.) (2016). *Costly and cute: Helpless infants and human evolution*. Albuquerque: University of New Mexico Press.

3. Steiner, P. (2019). Brain fuel utilization in the developing brain. *Annals of Nutrition and Metabolism, 75*(Suppl 1), 8–18; Knickmeyer, R. C., et al. (2008). A structural MRI study of human brain development from birth to 2 years. *Journal of Neuroscience, 28*(47), 12176–82.

4. Bjorklund, D. F. (1997). The role of immaturity in human development. *Psychological Bulletin, 122*(2), 153.

5. Bryce, E. (2019, Nov 9). How many calories can the brain burn by thinking? *Live Science*; Gilani, S. A. (2021). Can one burn calories just by thinking? Well, yes . . . a little bit. *Asian Journal of Allied Health Sciences (AJAHS), 4*(4); Kuzawa, C. W., et al. (2014). Metabolic costs and evolutionary implications of human brain development. *Proceedings of the National Academy of Sciences, 111(36)*, 13010–15; Kumar, A. (2020, Apr 27). The grandmaster diet: How to lose weight while barely moving. ESPN.com.

6. Kuzawa, C. W., et al. (2014).

7. Asrat, T., et al. (1991). Rate of recurrence of preterm premature rupture of membranes in consecutive pregnancies. *American Journal of Obstetrics and Gynecology, 165*(4), 1111–15.

8. Lester, B. M., Salisbury, A. L., Hawes, K., Dansereau, L. M., Bigsby, R., Laptook, A., . . . & Padbury, J. F. (2016). 18-month follow-up of infants cared for in a single-family room neonatal intensive care unit. *Journal of Pediatrics, 177*, 84–89; Feldman, R., & Eidelman, A. I. (2003). Skin-to-skin contact (Kangaroo Care) accelerates autonomic and neurobehavioural maturation in preterm infants. *Developmental Medicine & Child Neurology, 45*(4), 274–81.

9. Brito, N. H., et al. (2021, Aug 20). Paid maternal leave is associated with infant brain function at 3-months of age. PsyArXiv. https://doi.org/10.31234/osf.io/t4zvn.

10. Montez, K., Thomson, S., & Shabo, V. (2020). An opportunity to promote health equity: National paid family and medical leave. *Pediatrics, 146*(3)(Sep), e20201122. https://doi.org/10.1542/peds.2020-1122.

11. Montez, K. (2019, Jun 13). Dr. Kimberly Montez: Paid family leave puts babies' lives first. *Winston-Salem Journal*.

12. Romeo, R. R., Leonard, J. A., Robinson, S. T., West, M. R., Mackey, A. P., Rowe, M. L., & Gabrieli, J. D. (2018). Beyond the 30-million-word gap: Children's conversational exposure is associated with language-related brain function. *Psychological Science, 29*(5), 700–710.

13. Leung, C. Y. Y., Hernandez, M. W., & Suskind, D. L. (2020). Enriching home language environment among families from low-SES backgrounds: A randomized controlled trial of a home visiting curriculum. *Early Childhood Research Quarterly, 50*, 24–35.

14. Denworth, L. (2019, Apr 10). Hyperscans show how brains sync as people interact. *Scientific American*.

15. Nummenmaa, L., Putkinen, V., & Sams, M. (2021). Social pleasures of music. *Current Opinion in Behavioral Sciences, 39*, 196–202; Nummenmaa, L., Lahnakoski, J. H., & Glerean, E. (2018). Sharing the social world via intersubject neural synchronisation. *Current Opinion in Psychology, 24*, 7–14.

16. Stephens, G. J., Silbert, L. J., & Hasson, U. (2010). Speaker–listener neural coupling underlies successful communication. *Proceedings of the National Academy of Sciences, 107*(32), 14425–30.

17. Piazza, E. A., Hasenfratz, L., Hasson, U., & Lew-Williams, C. (2020). Infant and adult brains are coupled to the dynamics of natural communication. *Psychological Science, 31*(1), 6–17. https://doi.org/10.1177/0956797619878698.

18. Princeton University. Baby and adult brains sync during play, Princeton Baby Lab finds. Press release, Jan 9, 2020.

19. Romeo, R. R., et al. (2018).

20. Romeo, R. R., et al. (2021). Neuroplasticity associated with changes in conversational turn-taking following a family-based intervention. *Developmental cognitive neuroscience,* 100967; Gabrieli quoted in Anne Trafton (2018, Feb 13). Back-and-forth exchanges boost children's brain response to language. *MIT News.*

21. Leung, C. Y. Y., Hernandez, M. W., & Suskind, D. L. (2020).

22. List, J. A., Pernaudet, J., & Suskind, D. L. (2021). Shifting parental beliefs about child development to foster parental investments and improve school readiness outcomes. *Nature Communications, 12*(5765). https://doi.org/10.1038/s41467-021 -25964-y.

23. Siegler, A., & Zunkel, E. (2020, Jul). Rethinking federal bail advocacy to change the culture of detention. *Champion* magazine, National Association of Criminal Defense Lawyers, NACDL.org.

24. Rawls, J. (1971). *A theory of justice.* Cambridge, MA: Belknap Press of Harvard University.

25. Alexander, M. (2018, Oct 29). What if we're all coming back? *New York Times.*

CHAPTER FIVE

1. Roosevelt, E. Voice of America broadcast, Nov 11, 1951.

2. Kim, P., et al. (2010). The plasticity of human maternal brain: Longitudinal changes in brain anatomy during the early postpartum period. *Behavioral Neuroscience, 124*(5), 695.

3. Rigo, P., et al. (2019). Specific maternal brain responses to their own child's face: An fMRI meta-analysis. *Developmental Review, 51*(Mar), 58–69.

4. Gettler, L. T., et al. (2011). Longitudinal evidence that fatherhood decreases testosterone in human males. *Proceedings of the National Academy of Sciences, 108*(39), 16194–99; Storey, A. E., Alloway, H., & and Walsh, C. J. (2020). Dads: Progress in understanding the neuroendocrine basis of human fathering behavior. *Hormones and Behavior, 119,* 104660.

5. Taylor, S. E. (2006). Tend and befriend: Biobehavioral bases of affiliation under stress. *Current Directions in Psychological Science, 15*(6), 273–77.

6. Centers for Disease Control (1999). Achievements in public health, 1900–1999: Control of infectious diseases. *MMWR Weekly, 48*(29), 621–29.

7. Wrigley, J. (1989). Do young children need intellectual stimulation? Experts' advice to parents, 1900–1985. *History of Education Quarterly, 29*(1), 41–75.

8. Watson, J. B. (1928). *Psychological care of infant and child.* New York: W. W. Norton & Co.

9. Centers for Disease Control (1999).

10. Lareau, A. (2003). *Unequal childhoods: Race, class, and family life* (pp. 429–448). Berkeley: University of California Press; Lareau, A., et al. (2011). Ch. 15: Unequal

childhoods in context (pp. 333–41). In *Unequal Childhoods* (2nd ed.). Berkeley: University of California Press.

11. Ishizuka, P. (2019). Social class, gender, and contemporary parenting standards in the United States: Evidence from a national survey experiment. *Social Forces, 98*(1), 31–58.

12. Medina, J., Benner, K., & Taylor, K. (2019, Mar 12). Actresses, business leaders and other wealthy parents charged in U.S. college entry fraud. *New York Times.*

13. Leung, C. Y. Y., & Suskind, D. L. (2020). What parents know matters: Parental knowledge at birth predicts caregiving behaviors at 9 months. *Journal of Pediatrics, 221*, 72–80. https://doi.org/10.1016/j.jpeds.2019.12.021.

14. List, J. A., Pernaudet, J., & Suskind, D. L (2021). Shifting parental beliefs about child development to foster parental investments and improve school readiness outcomes. *Nature Communications, 12*(5765). https://doi.org/10.1038/s41467-021-25964-y.

15. ZERO TO THREE. (2016, Jun 6). *National parent survey overview and key insights.* See also, Roberts, M. Y., et al. (2019). Association of parent training with child language development: A systematic review and meta-analysis. *JAMA Pediatrics, 173*(7), 671–80.

16. Leung, C. Y. Y., & Suskind, D. L. (2020).

17. List, J. A., et al. (2021).

18. Leaper, C., Farkas, T., & Brown, C. S. (2012). Adolescent girls' experiences and gender-related beliefs in relation to their motivation in math/science and English. *Journal of Youth and Adolescence, 41*(3), 268–82; Beilock, S. L., et al. (2010). Female teachers' math anxiety affects girls' math achievement. *Proceedings of the National Academy of Sciences, 107*(5), 1860–63.

19. Berkowitz, T., et al. (2015). Math at home adds up to achievement in school. *Science, 350*(6257), 196–98.

20. Livingston, G. (2018, Sep 24). Stay-at-home moms and dads account for about one-in-five U.S. parents. Pew Research Center analysis of 2017 Current Population Survey Annual Social and Economic Supplements.

21. Gallup Poll, Aug 1–14, 2019. Record high 56% of U.S. women prefer working to homemaking.

22. Buteau, M. (2020, Dec 8). Helicopter mom vs. Jimmy Buffett dad. *New York Times.*

CHAPTER SIX

1. Mullainathan, S., & Shafir, E. (2014). *Scarcity: Why having too little means too much* (p. 41). New York: Henry Holt and Company.

2. Illinois State Board of Education. 2018–2019 Illinois Kindergarten Individual Development Survey (KIDS) Report: A look at kindergarten readiness; Overdeck Family Foundation (2020, Mar 4). The road to readiness: The precursors and practices that predict school readiness and later school success; Child Trends (2020, Apr 6). Comparing the national outcome measure of healthy and ready to learn with other well-being and school readiness measures; Blair, C., & Raver, C. C. (2015). School readiness and self-regulation: A developmental psychobiological approach. *Annual Review of Psychology, 66*, 711–31.

3. Williams, P. G., & Lerner, M. A. (2019). AAP policy on school readiness. *Pediatrics, 144*(2), e20191766. DOI: https://doi.org/10.1542/peds.2019-1766.

4. Child Trends (2020, Apr 6).

5. Illinois State Board of Education. 2018–2019 Illinois Kindergarten Individual Development Survey (KIDS) Report.

6. Regenstein, E. (2019, Feb). Why the K–12 world hasn't embraced early learning. *Foresight Law + Policy.*

7. Tamis-LeMonda, C. S., Luo, R., McFadden, K. E., Bandel, E. T., & Vallotton, C. (2019). Early home learning environment predicts children's 5th grade academic skills. *Applied Developmental Science, 23*(2), 153–69.

8. Duncan, G. J., et al. (2007). School readiness and later achievement. *Developmental Psychology, 43*(6), 1428; Engle, P. L., & Black, M. M. (2008). The effect of poverty on child development and educational outcomes. *Annals of the New York Academy of Sciences, 1136*, 243.

9. Sawhill, I., Winship, S., & Grannis, K. S. (2012, Sep 20). Pathways to the middle class: Balancing personal and public responsibilities. Brookings Institution.

10. National Scientific Council on the Developing Child (2020). *Connecting the brain to the rest of the body: Early childhood development and lifelong health are deeply intertwined.* Working Paper No. 15.

11. Mendelsohn, A. L., & Klass, P. (2018). Early language exposure and middle school language and IQ: Implications for primary prevention. *Pediatrics, 142*(4); Uccelli, P., et al. Children's early decontextualized talk predicts academic language proficiency in midadolescence. *Child Development, 90*(5), 1650–63; Demir, Ö. E., et al. (2015). Vocabulary, syntax, and narrative development in typically developing children and children with early unilateral brain injury: Early parental talk about the "there-and-then" matters. *Developmental Psychology, 51*(2), 161.

12. Gilkerson, J., et al. (2018). Language experience in the second year of life and language outcomes in late childhood. *Pediatrics, 142*(4).

13. Gilkerson, J., et al. (2017). Mapping the early language environment using all-day recordings and automated analysis. *American Journal of Speech-Language Pathology, 26*(2): 248–65.

14. Lino, M., et. al. (2017). *Expenditures on children by families.* U.S. Department of Agriculture Center for Nutrition Policy and Promotion. Miscellaneous report no. 1528-2015. Washington, DC: GPO.

15. Pew Research Center (2016, Oct). The future of work.

16. Ballentine, K. L., Goodkind, S., & Shook, J. (2020). From scarcity to investment: The range of strategies used by low-income parents with "good" low-wage jobs. *Families in Society, 101*(3), 260–74.

17. Abraham, K. G., et al. (2018). Measuring the gig economy: Current knowledge and open issues. No. 24950, NBER Working Papers; Board of Governors of the Federal Reserve System (2020, May). Report on the economic well-being of U.S. households in 2019.

18. Duszyriski, M. Gig economy: Definition, statistics & trends 2021 update. https://zety.com/blog/gig-economy-statistics.

19. Puff, J., and Renk, K. (2014). Relationships among parents' economic stress, parenting, and young children's behavior problems. *Child Psychiatry & Human Development, 45*(6), 712–27.

20. Malik, R., Hamm, K., Schochet, L., Novoa, C., Workman, S., & Jessen-Howard, S. (2018). America's child care deserts in 2018. Center for American Progress, pp. 3–4; National Institute of Child Health and Human Development (2006); Inside Early Talk (2021). LENA.
21. Inside Early Talk (2021).
22. Kartushina, N., et al. (2021, Mar 5). Covid-19 first lockdown as a unique window into language acquisition: What you do (with your child) matters. https://doi.org /10.31234/osf.io/5ejwu.
23. Roberts, A. M., Gallagher, K. C., Sarver, S. L., & Daro, A. M. (2018, Dec). Early childhood teacher turnover in Nebraska. Buffett Early Childhood Institute, University of Nebraska.
24. Frank, M. C., Braginsky, M., Yurovsky, D., & Marchman, V. A. (2021). *Variability and consistency in early language learning: The Wordbank Project.* Cambridge, MA: MIT Press.
25. Glass, J., Simon, R. W., & Andersson, M. A. (2016). Parenthood and happiness: Effects of work-family reconciliation policies in 22 OECD countries. *American Journal of Sociology, 122*(3), 886–929.

CHAPTER SEVEN

1. Withers, B. (1972). "Lean on Me." On *Still Bill.* Los Angeles: Sussex Records.
2. Mischel, W., Ebbesen, E. B., & Raskoff Zeiss, A. (1972). Cognitive and attentional mechanisms in delay of gratification. *Journal of Personality and Social Psychology, 21*(2), 204.
3. Shoda, Y., Mischel, W., & Peake, P. K. (1990). Predicting adolescent cognitive and self-regulatory competencies from preschool delay of gratification: Identifying diagnostic conditions. *Developmental Psychology, 26*(6), 978; Schlam, T. R., Wilson, N. L., Shoda, Y., Mischel, W., & Ayduk, O. (2013). Preschoolers' delay of gratification predicts their body mass 30 years later. *Journal of Pediatrics, 162*(1), 90–93.
4. Watts, T. W., Duncan, G. J., & Quan, H. (2018). Revisiting the marshmallow test: A conceptual replication investigating links between early delay of gratification and later outcomes. *Psychological Science, 29*(7), 1159–77.
5. Hughes, C., & Devine, R. T. (2019). For better or for worse? Positive and negative parental influences on young children's executive function. *Child Development, 90*(2), 593–609; Diamond, A., & Lee, K. (2011). Interventions shown to aid executive function development in children 4 to 12 years old. *Science, 333*(6045), 959–64.
6. Hughes and Devine (2019).
7. List, J. A., Pernaudet, J., & Suskind, D. L. (2021). Shifting parental beliefs about child development to foster parental investments and improve school readiness outcomes. *Nature Communications, 12*(5765). https://doi.org/10.1038/s41467 -021-25964-y.
8. Shonkoff, J. P., et al. (2010). *Persistent fear and anxiety can affect young children's learning and development.* National Scientific Council on the Developing Child, Working Paper No. 9; Hughes and Devine (2019); Shonkoff, J. P., et al. (2012). The lifelong effects of early childhood adversity and toxic stress. *Pediatrics, 129*(1), e232–e246.

9. National Scientific Council on the Developing Child (2007). *Excessive stress disrupts the architecture of the developing brain.* Working Paper No. 3. Cambridge, MA: Center on the Developing Child, Harvard University. http://developingchild .harvard.edu/index.php/resources/ reports_and_working_papers/working _papers/wp3/; Sandstrom, H., & Huerta, S. (2013, Sep). The negative effects of instability on child development: A research synthesis. The Urban Institute. https://www.urban.org/sites/default/files/publication /32706/412899-The -Negative-Effects-of-Instability-on-Child-Development-A-Research-Synthesis.pdf.

10. Cortés Pascual, A., Moyano Muñoz, N., & Quilez Robres, A. (2019). The relationship between executive functions and academic performance in primary education: Review and meta-analysis. *Frontiers in Psychology, 10,* 1582.

11. Blair, C., & Cybele Raver, C. (2015). School readiness and self-regulation: A developmental psychobiological approach. *Annual Review of Psychology, 66,* 711–31; Lin, H-L., Lawrence, F. R., & Gorrell, J. Kindergarten teachers' views of children's readiness for school. *Early Childhood Research Quarterly, 18*(2), 225–37.

12. Fox, N.A., & Shonkoff, J. P. (2012). How persistent fear and anxiety can affect young children's learning, behaviour and health. *Social and economic costs of violence: Workshop summary.* Washington, DC: National Academies Press.

13. Desmond, M., et al. (2013). Evicting children. *Social Forces, 92*(1), 303–27.

14. Centers for Disease Control and Prevention, Violence Prevention/Injury Center. Preventing adverse childhood experiences.

15. Harris, N. B. (2018). *The deepest well: Healing the long-term effects of childhood adversity.* Boston: Houghton Mifflin Harcourt.

16. Harris (2018); Pierce, L. J., et al. (2019). Association of perceived maternal stress during the perinatal period with electroencephalography patterns in 2-month-old infants. *JAMA Pediatrics, 173*(6), 561–70.

17. Palma-Gudiel, H., et al. (2015). Maternal psychosocial stress during pregnancy alters the epigenetic signature of the glucocorticoid receptor gene promoter in their offspring: A meta-analysis. *Epigenetics, 10*(10), 893–902.

18. Harvard Kennedy School, Government Performance Lab. Chicago, IL homelessness services performance improvement. https://govlab.hks.harvard.edu /chicago-il-homelessness-services-performance-improvement.

19. Lawson, G. M., et al. (2016). Socioeconomic status and the development of executive function: Behavioral and neuroscience approaches. In J. A. Griffin, P. McCardle, & L. S. Freund (Eds.), *Executive function in preschool-age children: Integrating measurement, neurodevelopment, and translational research* (pp. 259–278). American Psychological Association. https://doi.org/10.1037/14797-012.

20. National Scientific Council on the Developing Child (2010). *Persistent fear and anxiety can affect young children's learning and development.* Working Paper No. 9.

21. Desmond, M., & Kimbro, R. T. (2015). Eviction's fallout: Housing, hardship, and health. *Social Forces, 94*(1), 295–324.

22. Haelle, T. (2020, Apr 17). Postpartum depression can be dangerous. Here's how to recognize it and seek treatment. *New York Times.*

23. American Psychiatric Association. What is postpartum depression? https://www .psychiatry.org/patients-families/postpartum-depression/what-is-postpartum -depression.

24. Mental Health America (2021). The state of mental health in America; Social Solutions (n.d.). Top 5 barriers to mental healthcare access.

25. USA FACTS (2021, Jun 9). Over one-third of Americans live in areas lacking mental health professionals. https://usafacts.org/articles/over-one-third-of-americans-live -in-areas-lacking-mental-health-professionals/.
26. Andersson, M. A., Garcia, M. A., & Glass, J. (2021). Work-family reconciliation and children's well-being disparities across OECD countries. *Social Forces, 100*(2), 794–820.

CHAPTER EIGHT

1. Martin, W. P. (2004). *The best liberal quotes ever: Why the left is right* (p. 173). New York: Sourcebooks.
2. Yockel, M. (2007, May). 100 years: The riots of 1968. *Baltimore Magazine*; Maryland Crime Investigating Commission (1968). *A report of the Baltimore civil disturbance of April 6 to April 11, 1968.*
3. Gerald Grant, G. (1965, Apr 18). The vanishing lunch. *Washington Post, Times Herald*, E1.
4. King Jr., M. L. (1963, Apr 19). Letter from Birmingham jail.
5. Kurlansky, M. (2004). *1968: The year that rocked the world.* New York: Random House.
6. Brown, D. L. (2018, Apr 11). The Fair Housing Act was languishing in Congress. Then Martin Luther King Jr. was killed. *Washington Post*; The Civil Rights Act of 1968. Bullock Texas State History Museum. Austin, TX.
7. Mondale, W. W. (2010). *The good fight: A life in liberal politics.* New York: Scribner.
8. This account of the making of the CCDA is drawn from: Roth, W. (1976). The politics of daycare: The Comprehensive Child Development Act of 1971. Discussion Papers 369–76. Department of Health, Education, and Welfare, Washington, DC; Mondale (2010); Karch, A. (2013). *Early Start: Preschool policies in the United States.* Ann Arbor: University of Michigan Press; Badger, E. (2014, Jun 23). That one time America almost got universal child care. *Washington Post.*
9. Zigler, E., et al. (2009). *The tragedy of child care in America.* New Haven, CT: Yale University Press.
10. A new chance for children (1971, Aug 4). *Washington Post* editorial.
11. Statement announcing the establishment of the Office of Child Development, President Richard Nixon, April 9, 1969; American Presidency Project, UC Santa Barbara (1968, Aug 5). Republican Party platform of 1968.
12. California Newsreel. (2017, Jan 3). The veto that killed childcare [Video]. The raising of America. https://www.raisingofamerica.org/veto-killed-childcare.
13. Conservative lobbyist Jeff Bell quoted in Rose, E. (2010), *The promise of preschool: From Head Start to universal pre-kindergarten.* New York: Oxford University Press.
14. The Abecedarian Project, https://abc.fpg.unc.edu/abecedarian-project.
15. Ramey, C. T., et al. (2000). Persistent effects of early childhood education on high-risk children and their mothers. *Applied Developmental Science, 4*(1), 2–14.
16. Campbell, F. A., et al. (2002). Early childhood education: Young adult outcomes from the Abecedarian Project. *Applied Developmental Science, 6*(1), 42–57.
17. Campbell, F. A., et al. (2012). Adult outcomes as a function of an early childhood educational program: An Abecedarian Project follow-up. *Developmental Psychology, 48*(4), 1033.

18. Barnett, W. S., & Masse, L. N. (2007). Comparative benefit–cost analysis of the Abecedarian program and its policy implications. *Economics of Education Review, 26*(1), 113–25; Ramey, C. T., et al. (2000).

19. Ewing-Nelson, C. (2021). Another 275,000 women left the labor force in January. National Women's Law Center, Feb 2021 fact sheet; Stevenson quoted in Grose, J. (2021, Feb 4). The primal scream: America's mothers are in crisis. *New York Times.*

20. Centers for Disease Control (2020, Dec 10). Covid-19 racial and ethnic health disparities; Oppel, R. Jr, Gebeloff, R., Lai, K. K. R., Wright, W., & Smith, M (2020, Jul 5). The fullest look yet at the racial inequities of coronavirus. *New York Times.*

21. Buchanan, L., Bai, Q., & Patel, J. K. (2020, Jul 3). Black Lives Matter may be the largest movement in U.S. history. *New York Times.*

22. Justice Policy Institute (2015, Feb 25). The right investment?

23. Baltimore City Schools. District overview. https://www.baltimorecityschools.org /district-overview.

24. Sonenstein, F. L. (2014). Introducing the well-being of adolescents in vulnerable environments study: methods and findings. *Journal of Adolescent Health 55*(6): S1–S3.

25. U.S. Dept. of Health and Human Services. Children's Bureau timeline. https://www .childwelfare.gov/more-tools-resources/resources-from-childrens-bureau/timeline1.

26. Ladd-Taylor, M. (1992). Why does Congress wish women and children to die? The rise and fall of public maternal and infant health care in the United States, 1921–1929. In V. Fildes, et al. (Eds.) *Women and children first: International maternal and infant welfare, 1870–1945* (pp. 121–32). Routledge Revivals; Barker, K. (1998). Women physicians and the gendered system of professions: An analysis of the Sheppard-Towner Act of 1921. *Work and Occupations, 25*(2), 229–55; Moehling, C. M., & Thomasson, M. A. (2012), *Saving babies: The contribution of Sheppard-Towner to the decline in infant mortality in the 1920s.* No. 17996, NBER Working Papers. National Bureau of Economic Research.

27. Lemons, J. S. (1969). The Sheppard-Towner act: Progressivism in the 1920s. *Journal of American History, 55*(4), 776–86.

28. *JAMA* (1922, Jun 10). Proceedings of the St. Louis Session. *Journal of the American Medical Association 78*(22): 1804–17. https://doi.org/10.1001/jama.1922.02640 760034015.

29. Infant mortality and African Americans (2018). U.S. Department of Health and Human Services, Office of Minority Health.

30. The story of Dr. Andrus is drawn primarily from: Walker, C. (2018, Oct 8). Ethel Percy Andrus: One woman who changed America. *AARP*; Secrest, A. (2018, Jan 5). A chicken coop: The unlikely birthplace of AARP. *AARP.*

31. Harrington, M. (1962). *The other America: Poverty in the United States.* [Reprinted by Simon & Schuster, New York, 1997.]

32. Congressional Research Service (2021, Apr 14). Poverty among the population aged 65 and over.

33. Walker (2018), p. 116.

34. U.S. Census Bureau. New estimates on America's families and living arrangements. Press release, Dec 2, 2020.

35. Author interview with Elizabeth DiLauro and ZERO TO THREE. Strolling Thunder successes and stories.

36. Early learning Multnomah, from vision to vote. www.preschoolforall.org; Miller, C. C. (2020, Nov 6). How an Oregon measure for universal preschool could be a national model. *New York Times.*
37. Quoted in Secrest, A. (2018).

CHAPTER NINE

1. Franklin, B., writing anonymously, (1735, Feb 4). *Pennsylvania Gazette.*
2. Fleming, P. (2018, Jun 5). *The Accidental Epidemiologist.* Blog, Bristol Health Partners.
3. Perkins, A. (2016, Aug 26). Back to sleep: The doctor who helped stem a cot death epidemic. *Guardian.*
4. Fleming, P. (2018).
5. McKelvie, G. (2016, Jan 30). Anne Diamond has learned to smile again after suffering decades of heartbreak. *Mirror*; University of Bristol. (2017, February 10). BBC Points West report into life-saving cot death research [Video]. Youtube. https://www.youtube.com/watch?v=iUb8L8t9Yys.
6. Garstang, J., & Pease, A. S. (2018). A United Kingdom perspective. In J. R. Duncan & R. W. Byard (Eds.) *SIDS sudden infant and early childhood death: The past, the present and the future.* Adelaide, South Australia: University of Adelaide Press; American SIDS Institute, Incidence.
7. Fleming, P. (2018).
8. Gomez, R. E. (2016). Sustaining the benefits of early childhood education experiences: A research overview. *Voices in Urban Education, 43,* 5–14.
9. Schuster, M. A., et al. (2000). Anticipatory guidance: What information do parents receive? What information do they want? *Archives of Pediatrics & Adolescent Medicine, 154*(12), 1191–98. https://doi.org/10.1001/archpedi.154.12.1191; Garg, A., et al. (2019, Nov 1). Screening and referral for low-income families' social determinants of health by US pediatricians. *Academic Pediatrics, 19*(8), 875–83. https://doi.org/10.1016/j.acap.2019.05.125.
10. Maternal and Child Health Bureau (2016, Apr 24). National survey of children's health. https://mchb.hrsa.gov/data/national-surveys; Osterman, M. J. K., & Martin, J. A. (2018). Timing and adequacy of prenatal care in the United States, 2016. *National Vital Statistics Reports, 67*(3), 1–14; Centers for Disease Control and Prevention, National Center for Health Statistics, National Vital Statistics System.
11. Daro, D., Dodge, K. A., & Haskins, R. (2019). Universal approaches to promoting healthy development. *The Future of Children, 29*(1), 3–16; Garner, A. S. (2016). Thinking developmentally: The next evolution in models of health. *Journal of Developmental & Behavioral Pediatrics, 37*(7), 579–84.
12. Roberts, M. Y., et al. (2019). Association of parent training with child language development: A systematic review and meta-analysis. *JAMA Pediatrics, 173*(7), 671–80; Ipsos MORI, The Royal Foundation of the Duke and Duchess of Cambridge. (2020, Nov 27). State of the nation: Understanding public attitudes to the early years. Better Care Network. https://bettercarenetwork.org/sites/default/files/2020-12/Ipsos-MORI-SON_report_FINAL_V2.4.pdf

13. Sandel, M., et al. (2018). Unstable housing and caregiver and child health in renter families. *Pediatrics, 141*(2). https://doi.org/10.1542/peds.2017-2199.

14. Leung, C. Y. Y., & Suskind, D. L. (2020). What parents know matters: Parental knowledge at birth predicts caregiving behaviors at 9 months. *Journal of Pediatrics, 221*, 72–80.

15. Marvasti, F. F., & Stafford, R. S. (2012). From "sick care" to health care: Reengineering prevention into the U.S. system. *New England Journal of Medicine, 367*(10), 889–91. PubMed Central, https://doi.org/10.1056/NEJMp1206230.

16. Jolly, P., et al. (2013). U.S. graduate medical education and physician specialty choice. *Academic Medicine, 88*(4), 468–74.

17. Dodson, N. A., et al. (2021). Pediatricians as child health advocates: The role of advocacy education. *Health Promotion Practice, 22*(1), 13–17. https://doi.org/10.1177/1524839920931494.

18. Briggs, R., personal communication with Dana Suskind, [June 2021].

19. Schuster, M. A., et al. (2000, Dec). Anticipatory guidance: What information do parents receive? What information do they want? *Archives of Pediatric & Adolescent Medicine, 154*(12): 1191–98. https://doi.org/10.1001/archpedi.154.12.1191.

20. Leung, C., et al. (2018). Improving education on child language and cognitive development in the primary care settings through a technology-based curriculum: A randomized controlled trial. *Pediatrics, 142*(1_MeetingAbstract), p. 777.

21. Roberts, M. Y., et al. (2019). Association of parent training with child language development: A systematic review and meta-analysis. *JAMA Pediatrics, 173*(7), 671–80. https://doi.org/10.1001/jamapediatrics.2019.1197.

22. Council on Early Childhood, High, P., & Klass, P. (2014). Literacy promotion: An essential component of primary care pediatric practice. *Pediatrics, 134*(2), 404–9. https://doi.org/10.1542/peds.2014-1384.

23. Bradley, E. H., & Taylor, L. A. (2013). *The American health care paradox: Why spending more is getting us less.* New York: Public Affairs.

24. Fineberg, H. V. Foreword to Bradley and Taylor (2013), p. ix.

25. United Nations Population Fund. Trends in maternal mortality: 2000 to 2017, Executive Summary. https://www.unfpa.org/resources/trends-maternal-mortality-2000-2017-executive-summary. Accessed Sep 2, 2021.

26. Reitsma, M. B., et al. (2021, Jun). Racial/ethnic disparities in COVID-19 exposure risk, testing, and cases at the subcounty level in California. *Health Affairs, 40*(6), 870–78.

27. Bradley and Taylor (2013).

28. Fairbrother, G., et al. (2015). Higher cost, but poorer outcomes: The US health disadvantage and implications for pediatrics. *Pediatrics, 135*(6), 961. https://doi.org/10.1542/peds.2014-3298; Bradley, E. H., & Taylor, L. A. (2011, Dec 9). Opinion: To fix health, help the poor. *New York Times.*

29. World Health Organization (WHO) (n.d.). Taking action on the social determinants of health.

30. Garner (2016).

31. Rachel's story is told on the Family Connects website. https://familyconnects.org/our-stories/.

32. Hishamshah, M., et al. (2010). Belief and practices of traditional post partum care among a rural community in Penang Malaysia. *Internet Journal of Third World Medicine, 9*(2).

33. Maternity care—Kraamzorg Het Zonnetje (English). https://kraamzorghetzonnetje.nl/eng/maternity-care/.

34. Pao, M. (2017, Mar 26). States give new parents baby boxes to encourage safe sleep habits. NPR.

35. Dodge, K. A., et al. (2013). Randomized controlled trial of universal postnatal nurse home visiting: Impact on emergency care. *Pediatrics, 132*(Suppl 2), S140–S46; Dodge, K. A., et al. (2019). Effect of a community agency-administered nurse home visitation program on program use and maternal and infant health outcomes: A randomized clinical trial. *JAMA Network Open* 2(11), e1914522–e1914522; Goodman, W. B., et al. (2021). Effect of a universal postpartum nurse home visiting program on child maltreatment and emergency medical care at 5 years of age: A randomized clinical trial. *JAMA Network Open, 4*(7), e2116024–e2116024.

36. Canfield, C. F., et al. (2020). Encouraging parent-child book sharing: Potential additive benefits of literacy promotion in health care and the community. *Early Childhood Research Quarterly, 50*, 221–29. Earlier studies: High, P., Hopmann, M., LaGasse, L., and Linn, H. (1998, May). Evaluation of a clinic-based program to promote book sharing and bedtime routines among low-income urban families with young children. *Archives of Pediatrics and Adolescent Medicine, 152*(5), 459–65; Needlman, R., Fried, L. E., Morley, D. S., Taylor, S., and Zuckerman, B. (1991, Aug). Clinic-based intervention to promote literacy. A pilot study. *American Journal of Diseases of Children, 145*(8), 881–84; Weitzman, C. C., Roy, L., Walls, T., and Tomlin, R. (2004). More evidence for reach out and read: A home-based study. *Pediatrics, 113*(5), 1248–253; Sanders, L. M., Gershon, T. D., Huffman, L. C., and Mendoza, F.S. (2000, Aug). Prescribing books for immigrant children: A pilot study to promote emergent literacy among the children of Hispanic immigrants. *Archives of Pediatrics and Adolescent Medicine, 154*(8), 771–77.

37. Connor Garbe, M., et al. *The impact of Reach Out and Read among new and returning patients.* (Working manuscript.)

38. HealthySteps. www.healthysteps.org.

39. Center for Health Care Strategies (2018, May). Profile: Expanding awareness and screening for ACEs in the Bronx: Montefiore Medical Group.

40. Sege, R., et al. (2015). Medical-legal strategies to improve infant health care: A randomized trial. *Pediatrics, 136*(1), 97–106.

41. George Kaiser Family Foundation. Birth Through Eight Strategy for Tulsa (BEST). www.gkff.org/what-we-do/birth-eight-strategy-tulsa/.

42. Ready for School, Ready for Life. www.getreadyguilford.org.

CHAPTER TEN

1. Thoreau, H. D. (1854). *Walden.* (p. 164).

2. Oster, E. (2019, May 21). End the plague of secret parenting. *Atlantic.*

3. Friedman, M. (1970, Sep 13). The social responsibility of business is to increase profits. *New York Times,* SM, 17.

4. Williams, J. C. (2020, May 11). The pandemic has exposed the fallacy of the "ideal worker." *Harvard Business Review*.

5. Hacker, J. S. (2019). *The great risk shift: The new economic insecurity and the decline of the American dream*. New York: Oxford University Press.

6. Benford, E., et al. (2020, Aug 7). The Covid-19 eviction crisis: An estimated 30–40 million people in America are at risk. Aspen Institute.

7. Hacker (2019), pp. xi, 3.

8. Pew Charitable Trusts (2015, Jan 29). The precarious state of family balance sheets.

9. Hacker (2019), p. xiv.

10. Tavernise, S. (2021, May 5). The U.S. birthrate has dropped again. The pandemic may be accelerating the decline. *New York Times*.

11. Hacker (2019), p. 83.

12. Fisher, J., & Johnson, N. (2019). *The two-income trap: Are two-earner households more financially vulnerable?* Working Papers 19-19. Center for Economic Studies, U.S. Census Bureau.

13. Wang, W. (2018, Oct 14). The majority of U.S. children still live in two-parent families. Institute for Family Studies.

14. Dalu, M. C., et al. (2018, Nov 13). Why aren't U.S. workers working? FRBSF Economic Letter.

15. Kleven, H., et al. (2019). Child penalties across countries: Evidence and explanations. *AEA Papers and Proceedings, 109*.

16. Paquette, D., & Craighill, P. M. (2015, Aug 6). The surprising number of parents scaling back at work to care for kids. *Washington Post*. Results of a *Washington Post* poll.

17. Association of American Medical Colleges. More women than men enrolled in U.S. medical schools in 2017. Press release, December 17, 2017.

18. Delaying: Stack, S. W., Jagsi, R., Biermann, J. S., et al. (2020). Childbearing decisions in residency: A multicenter survey of female residents. *Academic Medicine, 95(10)*, 1550–557. https://doi.org/10.1097/ACM.0000000000003549. Beliefs: Kin, C., Yang, R., Desai, P., Mueller, C., & Girod, S. (2018). Female trainees believe that having children will negatively impact their careers: Results of a quantitative survey of trainees at an academic medical center. *BMC Medical Education, 18(1)*, 260. https://doi.org/10.1186/s12909-018-1373-1; Shifflette, V., Hambright, S., Amos, J. D., Dunn. E., & Allo, M. (2018). The pregnant female surgical resident. *Advances in Medical Education and Practice, 9*, 365–69. https://doi.org/10.2147/AMEP.S140738. Ob-gyn: Hariton, E., Matthews, B., Burns, A., Akileswaran, C., & Berkowitz, L. R. (2018). Pregnancy and parental leave among obstetrics and gynecology residents: Results of a nationwide survey of program directors. *American Journal of Obstetrics & Gynecology, 219(2)*, 199.e1–199.e8.

19. Frank, E., Zhao, Z., Sen, S., & Guille, C. (2019). Gender disparities in work and parental status among early career physicians. *JAMA Network Open, 2(8)*, e198340. https://doi.org/10.1001/jamanetworkopen.2019.8340.

20. Quoted in Petersen, A. H. (2020, Nov 11). Culture study, *Substack*. https://annehelen.substack.com/

21. Schultz's background taken from *My Story*, Howard Schultz personal website.

22. Starbucks Commitment to Partners (2017, Sep 1). https://stories.starbucks.com/press/2017/starbucks-commitment-to-partners.

23. Schultz, H. (2020, Sep 11). A free market manifesto that changed the world, reconsidered. *New York Times.*

24. Ball, P. (2015, Aug 12). How lifestyle benefits impact workplace productivity. Care.com. https://benefits.care.com/betterbenefits.

25. U.S. Chamber of Commerce Foundation. The bedrock of American business: High-quality early childhood education. https://www.uschamberfoundation.org /early-childhood-education/the-business-case.

26. Society for Human Resources Management. 2019 state of the workplace.

27. Ross, M., & Bateman, N. (2019, Nov). Meet the low-wage workforce. Brookings.

28. Vogtman, J., & Schulman, K. (2016). Set up to fail: When low-wage work jeopardizes parents' and children's success. National Women's Law Center.

29. Vogtman & Schulman (2016), pp. 13–16. See also Barnum, M. (2018, Sep 26). Here's a list of studies showing that kids in poverty do better in school when their families have more money. Chalkbeat.

30. Cwiek, S. (2014, Jan 27). The middle class took off 100 years ago . . . Thanks to Henry Ford? *All things considered.* NPR.

31. Dean, A., & Auerbach, A. (2018, Jun 5). 96% of U.S. professionals say they need flexibility, but only 47% have it. *Harvard Business Review*; Fuller, J. B., & Raman, M. (2019). The caring company: How employers can help employees manage their caregiving responsibilities—while reducing costs and increasing productivity. Harvard Business School, Managing the Future of Work Project; Li, J., et al. (2014). Parents' nonstandard work schedules and child well-being: A critical review of the literature. *Journal of Primary Prevention, 35*(1), 53–73.

32. Schaffner, M. (2021). Parents' just-in-time work schedules are not working for babies: A policy brief. ZERO TO THREE.; Collins, C. (2020, Nov 11). The free market has failed U.S. working parents. *Harvard Business Review.*

33. North, A. (2021, Jul 31). The five-day workweek is dead. *Vox.*

34. Williams, J. C., et al. (2018, Mar). *Stable Scheduling Study: Health outcomes report.* WorkLife Law (University of California, Hastings College of the Law).

35. Cahusac, E., & Kanji, S. (2014). Giving up: How gendered organizational cultures push mothers out. *Gender, Work & Organization, 21*(1), 57–70.

36. Miller, C. C. (2019, May 15). Work in America is greedy. But it doesn't have to be. *New York Times.*

37. Houser, L., & Vartanian, T. P. (2012). *Pay matters: The positive economic impacts of paid family leave for families, businesses and the public.* Report of the Rutgers Center for Women and Work; Paid leave fact sheets, National Partnership for Women and Families; Bartel, A. P., et al. (2021). *The impact of paid family leave on employers: Evidence from New York.* No. 28672, NBER Working Papers. National Bureau of Economic Research.

38. Osborne, C., Boggs, R., & McKee, B. (2021, Aug). Importance of father involvement. Child and Family Research Partnership, LBJ School of Public Affairs, University of Texas at Austin; Bartel, A. P., et al. (2018). Paid family leave, fathers' leave-taking, and leave-sharing in dual-earner households. *Journal of Policy Analysis and Management, 37*(1), 10–37; Petts, R. J., Knoester, C., & Waldfogel, J. (2020). Fathers' paternity leave-taking and children's perceptions of father-child relationships in the United States. *Sex Roles, 82*(3), 173–88; Rossin-Slater, M., & Stearns, J. (2020). Time on with baby and time off from work. *Future of Children,*

30(2), 35–51; Choudhury, A. R., & Polachek, S. W. (2019). *The impact of paid family leave on the timing of infant vaccinations*. IZA Institute of Labor Economics; Popper, N. (2020, Apr 17). Paternity leave has long-lasting benefits. So why don't more American men take it? *New York Times*; National Partnership for Women and Families (2021). Fact sheet: Fathers need paid family and medical leave.

39. National Partnership for Women and Families (2021). Fact sheet; Johannson, E-A. (2010). *The effect of own and spousal parental leave on earnings*. Working Paper No. 2010:4. Institute of Labour Market Policy Evaluation; Petts, R. J., Carlson, D. L., & Knoester, C. (2020, Oct). If I [take] leave, will you stay? Paternity leave and relationship stability. *Journal of Social Policy, 49*(4), 829–49.

40. Aspen Institute. Non-traditional work: Portable benefits. https://www.aspen institute.org/programs/future-of-work/nontraditional-work.

41. Kendra Scott story drawn from: Q +A with Kendra Scott, *Austin Family Magazine*; Hawkins, L. (2021, Feb 1). Kendra Scott steps aside as CEO of her Austin company. *Austin American-Statesmen*; Foster, T. (2018, Winter). This entrepreneur hit rock bottom before building a billion-dollar jewelry empire (with only $500). *Inc.*; Secrets of Wealthy Women podcast (2019, Jul 17). Kendra Scott: Building a billion dollar jewelry brand. *Wall Street Journal*.

42. Beth Ley of Kendra Scott interview. (2019, Jun 9). In How Kendra Scott promotes a family-friendly workplace. *InFocus Texas*; Kendra Scott Family Fund, www .kendrascottfamilyfund.com.

43. Best Place for Working Parents. https://bestplace4workingparents.com.

44. North Carolina's program is Family Forward NC. https://familyforwardnc.com.

CHAPTER ELEVEN

1. Franklin D. Roosevelt's Committee on Economic Security. Child welfare in the economic security program. Unpublished CES studies, Social Security online.

2. Library of Congress version.

3. In addition to interviewing Michael and Keyonna multiple times, I reviewed the court documents for Michael's case.

4. Luby, J. L. (2015). Poverty's most insidious damage: The developing brain. *JAMA Pediatrics, 169*(9), 810–11.

5. Goldin, C., & Katz, L. F. (2010). *The race between education and technology*. Cambridge, MA: Harvard University Press.

6. Chaudry, A. (2016, Aug). The case for early education in the emerging economy. *Roosevelt Institute Report* (pp. 3–4); Chaudry, A., & Sandstrom, H. (2020, Fall). Child care and early education for infants and toddlers. *Future of Children, 30*(2), 1717.

7. Weissman, D. G., et al. (2021, Nov 30). Antipoverty programs mitigate socioeconomic disparities in brain structure and psychopathology among U.S. youths. PsyArXiv. doi.org/10.31234/osf.io/8nhej.

8. Chaudry, A., et al. (2021). *Cradle to kindergarten: A new plan to combat inequality*. New York: Russell Sage Foundation.

9. The top five effective state policies are expanded income eligibility for health insurance, reduced administrative burden for SNAP, paid family leave, state minimum wage, and state earned income tax credit. The top six effective state

strategies are comprehensive screening and referral programs (like Family Connects), childcare subsidies, group prenatal care, evidence-based home-visiting programs, Early Head Start, and Early Intervention services. The complete road map is available at https://pn3policy.org/pn-3-state-policy-roadmap/complete-roadmap/.

10. Another example is Learning Policy Institute (2021). Building a national early childhood education system that works.

11. Chaudry et al. *Cradle to Kindergarten*, p. 137.

12. National Partnership for Women and Families (2018, Sep). Fact sheet: Paid family and medical leave: Good for business.

13. NICHD (2006).

14. Butrymowicz, S., & Mader, J. (2016, Mar 20). How the military created the best child care system in the nation. Hechinger Report; Covert, B. (2017, Jun 16). The U.S. already has a high-quality, universal childcare program—in the military. Think Progress; Allen, General J. R., & Lyles, General L. (2021, Aug 17). Why the child care crisis is a national security issue. *The Hill*.

15. Lucas, M. A. (2016, Apr 21). How the military went from having childcare in Quonset huts and stables to being the "premier" system—and why don't more people study it? Hechinger Report.

16. Lucas (2016), Hechinger Report.

17. Whitebook, M. (1989). *Who cares? Child care teachers and the quality of care in America. Final Report, National Child Care Staffing Study*. Oakland, CA: Child Care Employee Project.

18. Covert (2017).

19. Quoted in Covert (2017).

20. Child Care Aware of America (2013). We can do better: Child Care Aware of America's ranking of child care center regulations and oversight, 2013 update.

21. Helen Blank of the National Women's Law Center, quoted in Covert (2017).

22. Miller, C. C. (2021, Oct 6). How other nations pay for child care. The U.S. is an outlier. *New York Times*.

23. Emerson, S. (2018, Dec 27). California's First 5 programs evolve as smoking declines and tobacco taxes go away. *Daily Bulletin*, Southern California News Group.

24. There are twenty-two tribal communities in New Mexico, but they also collaborate with Fort Sill Apache Nation in Oklahoma, so some sources say twenty-three.

25. Annie E. Casey Foundation. (2021, Jun 21). 2021 Kids count data book: 2021 state trends in child well-being.

26. Families Valued. https://www.familiesvalued.org.

27. Anderson, R. (2021, Jan). Returning to work: Three policy steps to strengthen family life during the pandemic recovery. Center for Public Justice.

EPILOGUE

1. Douglass, F. *The North Star*. Frederick Douglass Papers at the Library of Congress.

• ACKNOWLEDGMENTS •

They say it takes a village to raise a child. Well, the same could be said for writing a book. I'm indebted to the many people who believed in these ideas and gave their time, wisdom, and guidance.

First and foremost, my deepest thanks belong to my writing partner, Lydia Denworth. Lydia and I share that ideal and rare partnership where two people are wholly aligned in vision but bring complementary skill sets—her journalistic restraint and my love of exclamation points, for example! Yet what made Lydia the ideal writing partner wasn't her journalistic excellence or scientific expertise but her passion for stories that can change the world. There is no one else whom I would have wanted to write this book with—I am forever grateful.

To the families and individuals whose stories fill the pages of this book, I cannot thank you enough. You let me into your lives and shared your most vulnerable moments. I will forever be in awe of your strength and moved by your boundless love. Thank you to Keyonna, Michael, Mariah, Sabrina, Randy, Elise, Katherine, Jade, Gabrielle, Rachel Anderson, Jovanna Archuleta, Kimberly Montez, Hazim Hardeman, Ellen Clarke, Dani Levine, and Talia Berkowitz.

To the amazing experts, visionaries, friends, and family who generously offered their wisdom and advice: This book could never have been written if not for you. My deepest thanks to: Ajay Chaudry, Chris Speaker, Cynthia Osborne, Deyanira Hernandez, Jovanna Archuleta, Elizabeth Groginsky, Jennifer Glass, Natalie Tackitt, Patrick Ishizuka, Rahil Briggs, Sara Redington, Stephanie Doyle, Patsy Hampton, Molly Day, Lai-Lani Ovalles, Naadia Owens, Kimberly Noble, Rachel Romeo, Susan Levine, Linda Smith, Perri Klass, Comer Yates, Luis

Avila, Karen Pekow, Elliot Regenstein, Ralph Smith, Beth Rash-baum, Diana Suskind, Sydnie Suskind, and Roberta Zeff. And a very special thanks to Eric Schultz for believing in these ideas and being such a thoughtful sounding board and support.

I'm deeply indebted to my amazing colleagues and partners at the TMW Center, no one more so than the incomparable Katie Dealy, our chief operating officer. Katie's pragmatic and principled leadership transformed us from a group of mission-oriented individuals to a unified center, aligned in vision and goals. She ensures that families are, and always will be, at the heart of everything we do. Katie, I am thankful for our partnership and relish every one of our frozen-yogurt-with-sprinkles celebrations.

From gorgeous writing to exceptional strategy to requisite logistics, Liz Sablich, TMW's director of communications, does it all. She is the rare person who is brave enough to say when bad ideas aren't worth pursuing, creative enough to realize when good ideas could be made better, and gracious enough to approach both situations with empathy and understanding. Liz, you are a gift.

My immense thanks also extend to Yolie Flores, who is bringing these ideas beyond the pages of a book and into the hands and hearts of parents across the country. She approaches every problem with humility, wisdom, and a passion to serve children and families. Yolie, everyone tells me how lucky I am that you joined me on this journey. I couldn't agree more.

Jon Wenger is the "jack of all trades" on the Parent Nation team. His creativity helped shape our mission, and all those around him feel his passion for service. Jon, you are brilliant and hardworking, but also the thing you most aspire to be, kind, and all three traits benefit TMW every day.

Heidi Stevens is another irreplaceable part of the Parent Nation team. With humanity and empathy, Heidi works tirelessly to lift up the families and organizations working to build a parent nation. Moreover, Heidi, a born storyteller, brought an irresistible belief that stories don't end at the reader—they can and should be our best chance for social change. Heidi, I'm profoundly thankful for our longstanding friendship . . . not least because it lured you into this important work.

I am so fortunate in the rest of the colleagues who have joined me at the TMW Center. Beth Suskind has been a friend, collaborator, and creative force on every project. Snigdha Gupta, who certainly did not know what she was signing up for, has been a strategic and indispensable partner in everything from teaching to tech development. Julie Pernaudet and Christy Leung constantly remind me of the art in science and the magic in discovery. Jodi Savitt's financial savvy charts our future and her remarkable kindness buoys our spirits. Kristin Leffel was with me at the inception of this research. Although Kristin is no longer at TMW, her impact and ideas are felt every day in the culture she helped establish. Hannah Caldwell and Dru Brenner brought every ounce of their passion, excitement, and vision to Parent Nation. They spearheaded a project to bring the tenets of Parent Nation into medical education and were invaluable contributors to countless projects. I also want to thank the entire team at TMW: Steph Avalos Bock, Amanda Bezik, Zayra Flores-Ayffan, Kelsey Foreman, Iara Fuenmayor Rivas, Teja Garre, Caroline Gaudreau, Paige Gulling, Debbie Hawes, Imrul Huda, Jacob Justh, Andy Lewis, Arnoldo Muller-Molina, Michelle Saenz, Melissa Segovia, Diana Smith, Alicia Taylor, Mia Thompson, and Milagra Ward.

I am also fortunate to have benefited from the support, trust, and friendship of many others during my career. While there are far too many names to list, I want to thank my longtime collaborators, thought partners, and friends Sally McCrady and Jeanine Fahnestock at PNC Grow Up Great™. Sally, Jeanine, and their incredible team have been champions of my work. Their investment in our home visiting studies advanced our scientific understanding and led me to many of the powerful stories that fill these pages. Beyond its support of the TMW Center's work, PNC Grow Up Great has been steadfast in its support of early childhood for more than seventeen years. It has demonstrated the powerful impact corporations/businesses can have in the lives of young children. Mildred Oberkotter, Bruce Rosenfield, David Pierson, and the wonderful team at the Oberkotter Foundation saw Parent Nation as a way to continue the Oberkotter Foundation's long and impactful history of supporting children and families with hearing loss to reach the promise of their promise. With their backing, I was able to dream bigger and reach for goals that I truly believe can change the world for the better. Amy Newnam, Teresa Caraway, and the Hearing First team provided endless guidance and steadied me with their unwavering belief that change is possible. Caroline Pfohl, Rob Kaufold, and Rick White at the Hemera Foundation believed in me and in this work when I was just beginning to move beyond the walls of the operating room. I am profoundly thankful to them for taking a chance on a surgeon and her ideas and for their passion for serving children and families. Without the support of these colleagues and countless others, there is no chance for a true parent nation.

As I worked on *Parent Nation,* I had a first-rate team of publishing professionals and advisors. I extend my deepest thanks to my editor, Stephen Morrow. Stephen believed in the vision and need for a parent nation from the beginning and helped me to make it a reality. Grace Layer's "invisible labor" ensured that all things moved smoothly. Stephanie Cooper, Isabel DaSilva, and Amanda Walker showed amazing dedication and excellence in the publicity and marketing of *Parent Nation.* My amazing copy and production editors, Joy Simpkins and LeeAnn Pemberton, operated with a level of care and attention to detail that rivals that of any surgeon. Elizabeth Shreve and Deb Shapiro at Shreve Williams elevated the ideas of *Parent Nation* and promoted the book across the country. Alexandra Gordon, Michael Lebakken, Claire Saget, Amanda Reid, Yasmin Hariri, Meredith Hellman, Rebecca Calautti, MJ Deery, and the team of visionaries, creators, and strategists at Matters Unlimited took the stories from

Parent Nation and flawlessly turned them into a cohesive campaign. My thanks also extend to Frank William Miller Jr. and Jess Rotter, who designed a cover that I hope this book gets judged by! Finally, I am grateful to Ana Aĺza, Clarissa Donnelly-DeRoven, Shiri Gross, Shanika Gunaratna, Marissa Jones, and Luz Kloz for reading countless drafts and scouring endless academic articles.

I deeply believe that there is no better place to write a book than the University of Chicago—a place where science and ideas are nurtured and where a surgeon can go from the operating room to the theater of social science. The ideas in *Parent Nation* are very much a reflection of my patients, my colleagues, and the intellectual exchange that happens so readily at this university. I want to extend my thanks to Jeff Matthews, who provided the initial spark I needed (Jeff, I hope I remain a positive ROI!); Nishant Agrawal for his continued support; and Fuad Baroody for the inspiring partnership and friendship throughout the years. Thank you to my wonderful colleagues on the Pediatric Hearing Loss and Cochlear Implant team, who put up with a surgeon who ventures away all too often: Brittney Sprouse, Katie Murdaugh, Caitlin Egan, Michael Gluth, Megan Greenya, Michelle Havlik, Ted Imbery, Stacy Pleasant, Samantha Dixon, Katie Swail, Emily Trittschuh, Gary Rogers, Nelson Floresco, and Robin Mills. And thanks to the innumerable professors, students, and colleagues who lent me their time, insight, and inspiration as I wrote *Parent Nation*.

Finally, I want to thank my family. To my parents, who showed me that service is the highest calling and taught me what it is to be a parent. To John, my love, my loudest cheerleader, and my fiercest advocate. And to my children, my greatest source of joy and my raison d'être. When I look at each of you, I am reminded daily of my generation's duty to create for our children the world we wished we had for ourselves, to leave an inheritance of love, opportunity, and justice, to build a parent nation.

—D.S.

*   *   *

It has been deeply satisfying to help my friend Dana Suskind bring this important book into the world. The breadth of her vision, the depth of her optimism, and the passion and joy she brings to her work inspire me daily. I am glad she thought to call me to ask for advice on hiring a writer. I think I surprised us both when I said, "I'd do that." I am so glad I did, and I thank her for giving me the opportunity and for bringing me into her world.

I am also thankful to Stephen Morrow and everyone at Dutton for believing in this book and helping to make it a reality. And thank you to my agent, Dorian Karchmar, for being such a strong supporter of me and my work.

Thanks, as ever, to the friends who so thoughtfully serve as my sounding boards, especially Moira Bailey, Stephanie Holmes, Elizabeth Schwarz, Leah MacFarlane, and Suzanne Myers. This time around, I was particularly grateful for the wise counsel and support of Christine Kenneally, Anya Kamenetz, and Reem Kassis. Our weekly Zoom calls have been a gift.

Working on this book has made me more thankful than ever for my good fortune in having Ray and Joanne Denworth as my parents. They gave me a loving and strong start in life, which has served me well in good times and bad. Envisioning a parent nation was also a powerful reminder of the joys and struggles of my own parenting journey. Jacob, Matthew, and Alex, I am thankful to be your mother. Thank you for your love, good humor, and patience. I do my best! I hope this book helps to make the world a better place for parents when it is your turn. Finally, thank you to Mark, my co-parent, for all that you do for our family, and all that you do for me.

—L.D.

• INDEX •

• ABOUT THE AUTHORS •

DANA SUSKIND, MD, is founder and co-director of the TMW Center for Early Learning + Public Health, director of the Pediatric Cochlear Implant Program, and professor of surgery and pediatrics at the University of Chicago. Dr. Suskind is the author of over forty-five scientific publications and *Thirty Million Words: Building a Child's Brain*. She is a member of the American Academy of Pediatrics and a fellow for the Council on Early Childhood. Her work has been profiled by numerous media outlets, including *The New York Times*, *The Economist*, *Forbes*, NPR, and *Freakonomics*.

LYDIA DENWORTH is an award-winning science writer and a contributing editor at *Scientific American*. She is the author of several books of popular science, including *Friendship: The Evolution, Biology and Extraordinary Power of Life's Fundamental Bond* and *I Can Hear You Whisper: An Intimate Journey through the Science of Sound and Language*.